MATHEMAGICAL BUFFET

Liong-shin Hahn

NTU Press
No. 1, Sec. 4, Roosevelt Rd., Taipei, Taiwan 106, R.O.C.
http://www.press.ntu.edu.tw
email: ntuprs@ntu.edu.tw

Mathemagical Buffet
By Liong-shin Hahn
© National Taiwan University Press 2013

GPN: 1010200022
ISBN: 978-986-03-5510-9

All rights reserved. No part of this book may be reproduced,
stored in a retrieval system, or transmitted, in any form
or by any means, electronic, mechanical, photocopying, recording,
or otherwise, without the written permission of
National Taiwan University Press.

Cover designed by Mula

To

My Special Friends

Professors

Tong-Shieng Rhai

Ju-Kwei Wang

and

My Grandsons

Kai-Jin and **Jin-Tai**

Contents

Preface		ix
1	Sums of Consecutive Integers	1
2	Galilean Ratios	7
3	The Pythagorean Theorem	11
	3.1 Proofs	11
	3.2 A Puzzle	17
	3.3 Pythagorean Triples	18
	3.4 Generalizations of the Pythagorean Theorem	20
4	Japanese Temple Mathematics	25
	4.1 Problems	25
	4.2 Solutions	27
5	Mind Reading Tricks	41
	5.1 Trick 1	41
	5.2 Trick 2	43
6	Magic Squares	47
	6.1 New Year Puzzle 2010	47
	6.2 Magic Squares	49
7	Fun with Areas	53
	7.1 Two Theorems of Newton	53
	7.2 A Charming Construction Problem	59
	7.3 A Generalization of the Simson Theorem	62
8	The Tower of Hanoi	67
9	Ladder Lotteries	71

10 Round Robin Competitions		**79**
11 Egyptian Fractions		**83**
12 The Ptolemy Theorem		**89**
	12.1 The Ptolemy Theorem	89
	12.2 Applications	90
	12.3 A Generalization of the Ptolemy Theorem	92
13 Convexity		**95**
	13.1 Introduction	95
	13.2 The Theorems of Jung and Helly	96
14 The Seven Bridges of Königberg		**99**
	14.1 Unicursal Figures	99
	14.2 Mazes	101
15 The Euler Formula		**105**
	15.1 The Euler Formula	105
	15.2 Regular Polyhedra	108
16 The Sperner Lemma		**111**
	16.1 The Sperner Lemma	111
	16.2 The Brouwer Fixed Point Theorem	117
	16.3 An Elementary Fixed Point Theorem	119
17 Lattice Points		**125**
	17.1 The Pick Theorem	125
	17.2 Lattice Equilateral Triangle	133
	17.3 Lattice Equiangular Polygons	135
	17.4 Lattice Regular Polygons	138
18 The Sums of Special Series		**139**
	18.1 The Sum of the Powers	139
	18.2 The Binomial Coefficients	141
	18.3 Faulhaber Polynomials	143
	18.4 The Sums of the Reciprocals of $S_p(n)$	150
	18.5 The Sums of Trigonometric Functions	153
19 The Morley Theorem		**157**

20 Angle Trisection — 163
- 20.1 Rules of Engagement 163
- 20.2 The Trisection Equation 167
- 20.3 Computations by Straightedge and Compass 169
- 20.4 Fields and their Extensions 172
- 20.5 Impossibility Proofs 176
- 20.6 Bending of the Rules 180
- 20.7 Regular Polygons 181
- 20.8 Regular 17-gon 185

21 Conics — 195

22 Primes — 203
- 22.1 Number of Primes 203
- 22.2 An Open Problem 205

23 Gaussian Integers — 209
- 23.1 Gaussian Primes 209
- 23.2 An Application to Real Primes 216

24 Calculus with Complex Numbers — 219

Appendix Determinants — 223
- A.1 Genesis 223
- A.2 Properties 232
- A.3 The Laplace Expansion Theorem 235

Preface

Throughout my many years of teaching at the university level, I frequently enjoyed giving stimulating little talks to secondary-school students (in the U.S. and Taiwan). This book contains detailed accounts of these talks. Because each topic is aimed at a particular grade level, chapters are independent and may be read in any order.

It is unlikely that more than a small fraction of our students will actually use mathematics in their careers. Therefore, we would do well to foster healthy attitudes and worthwhile habits of mind while they are in our care. In particular, I stress
(a) an appreciation of the beauty of mathematics,
(b) acquiring the habit of deep thinking,
(c) the ability to reason logically.
For this purpose, I can't think of a better choice than the revival of Euclidean geometry in our secondary school curriculum. Euclidean geometry contains a wealth of not only wonderful theorems for students to enjoy the beauty of mathematics, but also intriguing and challenging problems to lure students into deep thinking and explorations; not to mention that it provides a fertile ground for training in logical reasoning. In fact, I have heard many people in my generation and earlier who claimed that they enjoyed Euclidean geometry immensely, even though they were not good at algebra. As Albert Einstein said, "If Euclid failed to kindle your youthful enthusiasm, then you were not born to be a scientific thinker."

Mathematics books are not to be "read". They should be worked through with pencil and paper. Looking back on my own experience, I learned mathematics not by reading books, nor by attending lectures. I learned mathematics mainly by solving (or trying to solve) challenging problems; by trying to explore what happens when part of the assumption of the theorem is altered or deleted; by trying to find what it means in simple particular cases; by investigating the converse. In short, by playing around with problems. Furthermore, my experience convinces me that the crux of a great theorem lies often in a simple concrete special case. Consequently, in teaching, I try to emphasize the important particular cases rather than the most general case. And I try my best to expose the motivation behind each move; at the minimum, I try to avoid presenting solutions and proofs as beautiful but "static" finished artifacts.

I believe that mathematics textbooks should emphasize ideas; they should not be mere collections of the facts. My teaching motto is: "Don't try to teach everything. Always leave something for students to explore." Hence I tell my students, "I do the easy part and you do the hard part." Consequently, I am allergic to overweight textbooks trying to include everything. Does anyone really believe students are interested in reading 5-pound, 800-page textbooks with most of the pages filled with repetitions of simple routine drills, ad nauseam? I get the impression that authors of overweight textbooks are more concerned with encyclopedic coverage of topics at the expense of discussing how topics relate to each other. Consequently, students think that mathematics is just a collection of facts, facts to be remembered in order to pass multiple-choice tests. How tragic it is to starve millions of eager young minds by depriving them of being exposed to the beauty and excitement of mathematics!

This book goes against the current trend. I try to present something intriguing that does not yield to well-worn standard approaches, something that involves a spark of ingenuity. The book is now presented to be judged by readers. I cherish the hope that you will enjoy the feast in my *Mathemagical Buffet*.

L.-s. Hahn

Acknowledgment. Words can not express my appreciation for the help by Professor Ross Hansberger. I treasure his friendship dearly. It is my pleasure to express my gratitude for Ms. Tzu-ling Yu (the editor of *Honsberger Revisited*) and Mr. Yu-lin Wu (the editor of *Mathemagical Buffet*). Their efficiency and cooperation are beyond my expectation. In both cases, the books were published within one year of the submission of the manuscripts to the National Taiwan University Press. Finally, I am deeply indebted to my son, Shin-Yi. Without his help in solving my computer problems, I would not be able to write even a single book after my retirement.

Chapter 1

Sums of Consecutive Integers

Here is a scenario that could be enacted in a classroom, say for 7th graders, if not for 6th graders.

T: We know 6 can be expressed as the sum of three consecutive positive integers as $6 = 1+2+3$. Can you find other numbers that can be expressed as the sums of two or more consecutive positive integers?

S: That's easy. Just add 4 at the end of $1+2+3$; namely,
$$1+2+3+4 = 10; \quad \text{and} \quad 1+2+3+4+5 = 15, \text{ etc.}$$

T: Good. But you don't have to always start from 1.

S: Then drop 1 at the beginning to get
$$2+3 = 5, \quad 2+3+4 = 9, \quad 2+3+4+5 = 14, \text{ etc.}$$

S: We can go further by dropping 2 here. Of course dropping 2 from $2+3$ leaves only one term, and that's not good. But for others, we get
$$3+4 = 7, \quad 3+4+5 = 12, \text{ etc.}$$

T: So, do you think every number can be expressed as the sum of two or more consecutive positive integers?

S: No, 1 and 2 can't be. The smallest such number is $1+2 = 3$.

T: Very good. How about 4?

S: $1+2$ is too small, and $2+3$ is already too big. So 4 can not be expressed as the sum of two or more consecutive positive integers.

T: We have seen $1+2 = 3$, $2+3 = 5$, $3+4 = 7$.

1

S: And the next is $4+5=9$; followed by $5+6=11$.

S: I get it. Every odd positive integer can be expressed as the sum of two consecutive positive integers.

T: Excellent! But remember, 1 is also an odd positive integer.

S: O.K., except 1.

T: Perfect! Every odd positive integer greater than 1 can be expressed as the sum of two consecutive positive integers. How about the converse?

S: Converse is easy. The sum of two consecutive integers is obviously an odd integer because one of the two consecutive integers is even while the other is odd.

T: Good! That takes care of the sums of the two consecutive integers. We saw 6 can be expressed as the sum of three consecutive positive integers already. And clearly, $6 = 1 + 2 + 3$ is the smallest among such integers.

S: It is followed by $2+3+4=9$, $3+4+5=12$, $4+5+6=15$, etc.

T: Do you see a pattern here?

S: They are all multiples of 3, but not 3 itself.

T: Great! But can you prove it?

S: Well, the sum of 3-term arithmetic progression is 3 times the middle term, say m. Hence the sum must be $3m$.

T: Good. What you have shown is that if a number can be expressed as the sum of three consecutive integers, then it must be a multiple of 3. What about the converse? Can every multiple of 3 except 3 itself be expressed as the sum of three consecutive positive integers?

S: It's easy. If n is a multiple of 3, then we divide n by 3 to obtain the middle term m; i.e., $m = n/3$. And we can express n as

$$n = (m-1) + m + (m+1).$$

T: So what's wrong with $n = 3$?

S: If $n = 3$, then $m = 1$, and the first term $m - 1 = 0$ is not positive.

T: Good!

S: Similarly, the sum of 5 consecutive positive integers must be a multiple of 5; and conversely, a multiple of 5 can be expressed as the sum of 5 consecutive positive integers if it is not too small, say at least $1 + 2 + 3 + 4 + 5 = 15$.

T: Wonderful!

S: In fact, our argument works for any odd number of terms.

T: Very good. So what cases are still not covered?

S: The sums of four consecutive integers, six consecutive integers, etc.

T: What is the smallest among such numbers?

S: $1+2+3+4=10$. It's not a multiple of 4.

T: What is the average of these four integers?

S: The average is 2.5. It's not an integer.

T: And for 6 consecutive integers?

S: $1+2+3+4+5+6=21$, and the average in this case is 3.5; again not an integer.

S: I get it! Even though the average is not an integer, it is half of an (odd) integer. But because we have even number of terms, the sum turns out to be an integer.

T: Good point! Actually, we saw that the sum of two consecutive integers is always odd; so it can never be a multiple of 2. For any even number of consecutive integers, the average is always the midpoint between two consecutive positive integers, say m and $m+1$. That is, the average is $\frac{1}{2}\{m+(m+1)\} = \frac{2m+1}{2}$, and we have even number of terms, say $2k$. Therefore, the sum n must be

$$n = \frac{2m+1}{2} \cdot (2k) = (2m+1) \cdot k.$$

Recall: if we have odd number of terms, say $2k+1$ terms ($k \geq 1$), then the average, say m, is a positive integer, and so the sum n is

$$n = m \cdot (2k+1).$$

S: So regardless of the number of terms, in order for a positive integer n to be expressible as the sum of two or more consecutive positive integers, n must have an odd divisor.

T: Good observation! Now can we claim the converse that if a positive integer n has an odd divisor, say $2k+1$, then n can be expressed as the sum of consecutive positive integers?

S: I think so. If a positive integer n has an odd divisor, say $2k+1$, then the quotient $m = \frac{n}{2k+1}$ is a positive integer, and we can put k terms each before and after m to obtain

$$\begin{aligned} n &= \{(m-k)+(m-k+1)+\cdots+(m-1)\}+m \\ &\quad +\{(m+1)+(m+2)+\cdots+(m+k)\}. \end{aligned}$$

S: Something puzzles me. It appears that we always have odd number of terms. What happened to the cases of even number of terms?

T: Let's do some experiments. Try $10 = 2 \times 5 = 2 \cdot (2 \cdot 2 + 1)$.

S: To have a 5-term representation, 2 must be at the center, and we must have 2 terms each before and after 2; but two terms before 2 is 0, and that's not positive.

T: Never mind. Just write it down.

S: O.K., we have $10 = (0+1) + 2 + (3+4)$. But only positive integers counts, so we get
$$10 = 1 + 2 + 3 + 4.$$
We get 4 terms instead of 5!

T: Yes, we saw $3 = 0 + 1 + 2$ before. How about $14 = 2 \times 7 = 2 \cdot (2 \cdot 3 + 1)$ or $18 = 2 \times 9 = 2 \cdot (2 \cdot 4 + 1)$?

S: To have a 7-term representation with 2 at the center, we should have three terms each before and after 2:
$$14 = \{(-1) + 0 + 1\} + 2 + \{3 + 4 + 5\} = 2 + 3 + 4 + 5.$$
Similarly,
$$18 = \{(-2) + (-1) + 0 + 1\} + 2 + \{3 + 4 + 5 + 6\} = 3 + 4 + 5 + 6.$$
Again we get even number of terms.

T: Recall if n has an odd divisor, say $2k+1$, then there must be a positive integer m such that $n = m \cdot (2k+1)$. And we can write n as the sum of $2k+1$ consecutive integers with m at the middle:
$$\begin{aligned} n = & \{(m-k) + (m-k+1) + \cdots + (m-1)\} + m \\ & + \{(m+1) + (m+2) + \cdots + (m+k)\}. \end{aligned}$$

S: So when $m > k$, then we have odd number of terms; but when $m \leq k$, then in addition to 0, two terms that differ by signs (if any) cancel each other, so these terms are redundant, and deleting these odd number of terms, we obtain even number of terms.

T: Good! That implies that we can not have only one term left because 1 is not even. But how do you know there won't be a case with no term left?

S: Well, in order for all the (positive) terms to be cancelled, the average m must be 0 (or negative). But the middle term m, being the quotient of two positive integers, is always positive.

T: That's right! Hence for each odd divisor (of an integer n) greater than 1, we obtain one representation of n as the sum of two or more consecutive positive integers. For example, try 15.

S: Let's see 15 has odd divisors 3, 5 and 15 (that are greater than 1). And from $15 = 5 \times 3 = 5 \cdot (2 \times 1 + 1)$, we obtain
$$15 = 4 + 5 + 6.$$
From $15 = 3 \times 5 = 3 \cdot (2 \times 2 + 1)$, we obtain
$$15 = 1 + 2 + 3 + 4 + 5.$$
And finally, from $15 = 1 \times 15 = 1 \cdot (2 \times 7 + 1)$, we obtain
$$\begin{aligned} 15 &= \{(-6) + (-5) + (-4) + (-3) + (-2) + (-1) + 0\} + 1 \\ &\quad + \{2 + 3 + 4 + 5 + 6 + 7 + 8\} \\ &= 7 + 8. \end{aligned}$$
That is, 15 can be expressed in three ways as the sum of two or more consecutive positive integers.

T: However, we must make sure that two different odd divisors (of a positive integer n) will never give a same representation of n as the sum of two or more consecutive positive integers.

S: Well, given a representation (of a positive integer) as the sum of two or more consecutive positive integers, if the number of the terms is odd, then the corresponding odd divisor is the number of the terms. And if the number of the terms is even, then the average of the terms must be half of an odd integer. So the corresponding odd divisor is twice of the average of the terms. In either case, each representation determines the corresponding odd divisor uniquely, and the converse is obviously true. So two distinct odd divisors (of the given positive integer) can not give rise to a same representation.

T: Very good! Therefore, the only numbers that can not be expressed as the sum of two or more positive integers are numbers of the form 2^k ($k = 0, 1, 2, 3, \cdots$).

S: Because they are the only ones without any odd divisor (other than 1).

T: Good! We have a complete solution of this problem.

Exercise. Compose a classroom scenario with a topic of your own choice.

Chapter 2

Galilean Ratios

Can you find the relation between the equalities and Figure GLO/Left?

$$1 = 1^2$$
$$1 + 3 = 2^2$$
$$1 + 3 + 5 = 3^2$$
$$1 + 3 + 5 + 7 = 4^2$$
$$1 + 3 + 5 + 7 + 9 = 5^2$$
$$\cdots = \cdots.$$

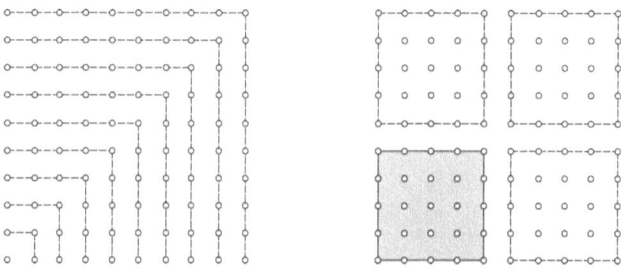

Figure GLO

Note that, because we have

$$1 + 3 + 5 + 7 + 9 = 5^2,$$

it follows that

$$\begin{aligned}
1 + 3 + 5 + 7 + 9 + 11 &= (1 + 3 + 5 + 7 + 9) + 11 \\
&= 5^2 + (2 \times 5 + 1) = 5^2 + 2 \times 5 + 1 \\
&= (5 + 1)^2 = 6^2.
\end{aligned}$$

7

In general, if
$$1 + 3 + \cdots + (2n-1) = n^2,$$
then
$$\begin{aligned} 1 + 3 + \cdots + (2n-1) + (2n+1) \\ &= \{1 + 3 + \cdots + (2n-1)\} + (2n+1) \\ &= n^2 + 2n + 1 = (n+1)^2. \end{aligned}$$

So, if the equality holds at some stage, then it holds also at the next stage. But it does hold for the first case ($n = 1$), so it must hold for the next case $n = 2$, and then $n = 3$, $n = 4$, etc.

By the way, from
$$(n+1)^2 = n^2 + (2n+1),$$
we notice that if $2n+1$ is a perfect square, then we have the so-called *Pythagorean triple*. (Can you express this identity by arranging dots as in Figure GLO/Left?) For example, if $2n+1 = 3^2$, then $n = 4$, and we obtain $5^2 = 4^2 + 3^2$. And if $2n+1 = 5^2$, then $n = 12$, and we have $13^2 = 12^2 + 5^2$.
In general, setting $k^2 = 2n+1$, and rewriting the identity above in terms of k, we obtain
$$\left(\frac{k^2+1}{2}\right)^2 = \left(\frac{k^2-1}{2}\right)^2 + k^2.$$
Thus $\{k, \frac{k^2-1}{2}, \frac{k^2+1}{2}\}$ is a Pythagorean triple for arbitrary odd integer $k \geq 3$.

Next, observe that
$$\frac{1}{3} = \frac{1+3}{5+7} = \frac{1+3+5}{7+9+11} = \frac{1+3+5+7}{9+11+13+15}$$
$$= \frac{1+3+5+7+9}{11+13+15+17+19} = \cdots$$

Exercise. Can you generalize? That is, can you find similar equalities?

From our discussion above, we have
$$\frac{1+3+5+7+9}{11+13+15+17+19}$$
$$= \frac{1+3+5+7+9}{(1+3+5+7+9+11+13+15+17+19) - (1+3+5+7+9)}$$
$$= \frac{5^2}{10^2 - 5^2} = \frac{5^2}{(2 \cdot 5)^2 - 5^2} = \frac{5^2}{5^2(2^2 - 1^2)} = \frac{1}{4-1} = \frac{1}{3}.$$

Galilean Ratios

Note carefully, in Figure GLO/Right, the solid square corresponds to the numerator, and the union of the remaining three squares corresponds to the denominator.

In general, once we have

$$1 + 3 + \cdots + (2n - 1) = n^2 \quad \text{for all positive integer } n,$$

then,

$$\begin{aligned}
&\frac{1 + 3 + \cdots + (2n - 1)}{(2n + 1) + \cdots + [2(2n) - 1]} \\
&= \frac{1 + 3 + \cdots + (2n - 1)}{\{1 + 3 + \cdots + [2(2n) - 1]\} - \{1 + 3 + \cdots + (2n - 1)\}} \\
&= \frac{n^2}{(2n)^2 - n^2} = \frac{n^2}{n^2(2^2 - 1^2)} = \frac{1}{4 - 1} = \frac{1}{3}.
\end{aligned}$$

Thus we have the Galilean ratios.

As for generalization, instead of cutting a square into four equal squares, what if we cut a square into 9 equal squares? Any other way of cutting?

Actually, there is another picturesque way to prove Galilean ratio:

$$\begin{aligned}
\frac{1}{3} &= \frac{1 + 3}{5 + 7} = \frac{1 + 3 + 5}{7 + 9 + 11} = \frac{1 + 3 + 5 + 7}{9 + 11 + 13 + 15} \\
&= \frac{1 + 3 + 5 + 7 + 9}{11 + 13 + 15 + 17 + 19} = \cdots
\end{aligned}$$

Simply look at Figure GLO1. Get it? Can you generalize?

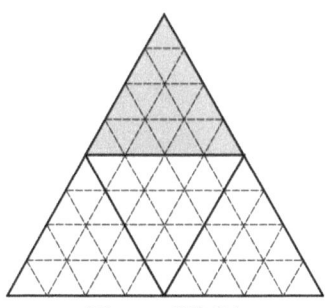

Figure GLO1

Exercise. How many equilateral triangles are there in Figure GLO1? Can you find a general formula?

Chapter 3

The Pythagorean Theorem

3.1 Proofs

The Pythagorean theorem is arguably the most important theorem in mathematics.

Theorem 1. Let a, b, c be the three side lengths of $\triangle ABC$ opposite the vertices A, B, C, respectively. Then
$$\angle C = 90° \iff a^2 + b^2 = c^2.$$

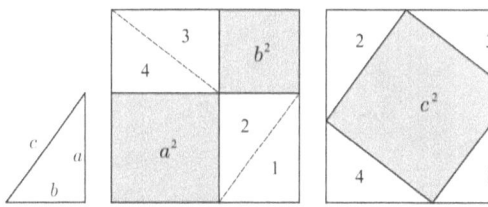

Figure PYG

Here is my favorite proof. Just compare the two squares in Figure PYG. After subtracting four triangles that are congruent to the given right triangle from each of the squares with the side length $a + b$, the remaining areas must be the same. For the square on the left, we are left with two squares, one with the side length a, the other b; while for the square on the right, we are left with one square (Check that the angles are $90°$!) with the side length c. So we must have
$$a^2 + b^2 = c^2.$$

Put it in another way, because we are subtracting four right triangles (each with the area $\frac{1}{2}ab$) from the square whose side length is $a+b$, the remaining area must be
$$(a+b)^2 - 4\left(\frac{ab}{2}\right) = (a^2 + 2ab + b^2) - 2ab = a^2 + b^2.$$

But in Figure PYG/Right, what remains is a square whose side length is c. Therefore,
$$a^2 + b^2 = c^2.$$
The square in Figure PYG/Left is just a geometric version of the algebraic formula
$$(a+b)^2 = a^2 + 2ab + b^2.$$
Some beginners tend to forget the middle term $2ab$ (in this formula), which is the sum of the areas of the two rectangles, but this picture hopefully will remind them why this term is indispensable.

As long as we are willing to use a little algebra, we can give yet another proof. (Figure PYG1) This time, the big square outside has the side length c, the small square in the middle has the side length $a - b$ (or $b - a$), and each of the four triangles has the area $\frac{1}{2}ab$. So
$$c^2 = (a-b)^2 + 4\left(\frac{ab}{2}\right) = (a^2 - 2ab + b^2) + 2ab = a^2 + b^2.$$

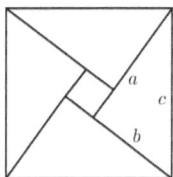

Figure PYG1

Here is a short proof using similarity. Let H be the feet of the perpendicular from vertex C to hypotenuse AB. (Figure PYG2/Right) Then it is easy to see that $\triangle ABC \sim \triangle CBH$. Therefore,
$$\frac{\overline{BC}}{\overline{BA}} = \frac{\overline{BH}}{\overline{BC}} \; ; \text{ i.e., } \overline{BC}^2 = \overline{BA} \cdot \overline{BH}.$$
Similarly, from $\triangle ABC \sim \triangle ACH$, we obtain
$$\overline{CA}^2 = \overline{BA} \cdot \overline{HA}.$$
Adding the last two equalities, we obtain
$$\overline{BC}^2 + \overline{CA}^2 = \overline{BA} \cdot (\overline{BH} + \overline{HA}) = \overline{AB}^2.$$

The Pythagorean Theorem

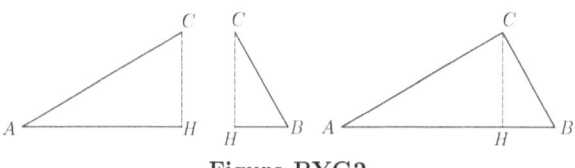

Figure PYG2

Actually, from Figure PYG2, we obtain the Pythagorean theorem instantly and elegantly if we are aware that, to each set of similar plane figures, there exists a (nonzero) constant k such that the area of each member in the set is given by

$$k \times \text{(the square of the corresponding length)}.$$

For example, all equilateral triangles are similar, and the area is given by

$$\frac{\sqrt{3}}{4} \times \text{(the side length)}^2;$$

i.e., $k = \frac{\sqrt{3}}{4}$ for equilateral triangles. For squares, the constant is 1 (corresponding to the side length), and for circles, the constant is π (corresponding to the radius); etc.

Now, in Figure PYG2, $\triangle ACH$, $\triangle CBH$, $\triangle ABC$ are all similar, so there must be a (nonzero) constant k (corresponding to the hypotenuse) such that

$$[ACH] = k \cdot \overline{CA}^2, \quad [CBH] = k \cdot \overline{BC}^2, \quad [ABC] = k \cdot \overline{AB}^2.$$

But we know
$$[ACH] + [CHB] = [ABC],$$

Therefore,
$$k \cdot \overline{CA}^2 + k \cdot \overline{BC}^2 = k \cdot \overline{AB}^2, \quad \overline{CA}^2 + \overline{BC}^2 = \overline{AB}^2.$$

Exercise. (Hippokrates) Find the sum of the areas of the two shaded regions where three semicircles are drawn on the three sides of a right triangle. (Figure PYG3)

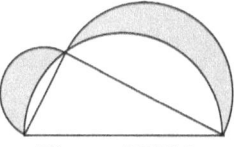

Figure PYG3

As for the converse part of the Pythagorean theorem, suppose a, b, c are the lengths of the three sides of $\triangle ABC$ that are opposite the vertices A, B, C, respectively, satisfying

$$a^2 + b^2 = c^2.$$

Draw a right triangle $A'B'C'$ as in Figure PYG4 such that

$$\overline{B'C'} = \overline{BC}, \quad \overline{C'A'} = \overline{CA}, \quad \text{and} \quad \angle A'C'B' = 90°.$$

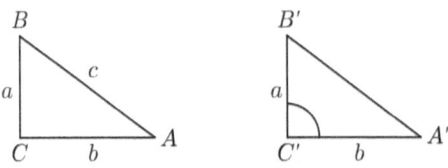

Figure PYG4

Then, by the necessary part of the Pythagorean theorem (that we have already established), we have

$$\begin{aligned} \overline{A'B'}^2 &= \overline{B'C'}^2 + \overline{C'A'}^2 = \overline{BC}^2 + \overline{CA}^2 \\ &= a^2 + b^2 = c^2 = \overline{AB}^2, \end{aligned}$$

implying that $\overline{A'B'} = \overline{AB}$. Therefore, $\triangle A'B'C' \cong \triangle ABC$ (SSS). It follows that $\angle ACB = \angle A'C'B' = 90°$.

Exercise. In Figure PYG5, O is a point on the extension of diameter AB; M is the center of the circle; R is the point on the circle such that $MR \perp AB$; OG is tangent to the circle; and H is the foot of the perpendicular from G to diameter AB. Because the hypotenuse is greater than a leg in any right triangle, we have

$$\overline{OR} > \overline{OM} > \overline{OG} > \overline{OH}.$$

Express these inequalities in terms of $a = \overline{OA}$ and $b = \overline{OB}$.

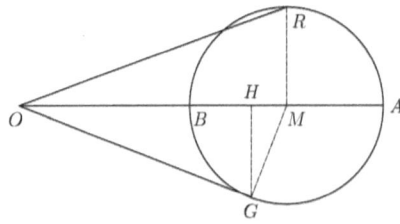

Figure PYG5

From the exercise above, we have

$$\sqrt{\frac{a^2 + b^2}{2}} > \frac{a+b}{2} > \sqrt{ab} > \frac{2ab}{a+b} \quad (a > b > 0).$$

The Pythagorean Theorem

The terms in the inequalities are called (in order) the *root mean square*, the *arithmetic mean*, the *geometric mean*, and the *harmonic mean*, of a and b, respectively. Note that each of these means are symmetric with respect to a and b (i.e., their values do not change even if we interchange a and b), therefore, the inequalities must hold for $0 < a < b$ too. (What if $a = b$?)

Note also that we have established these inequalities by geometry. Naturally, they can be established purely algebraically. For example,

$$\left(\frac{a+b}{2}\right)^2 - \left(\sqrt{ab}\right)^2 = \frac{a^2 + 2ab + b^2}{4} - ab$$

$$= \frac{a^2 - 2ab + b^2}{4} = \left(\frac{a-b}{2}\right)^2 \geq 0.$$

Therefore, we have the A.M.-G.M. inequality

$$\frac{a+b}{2} \geq \sqrt{ab} \quad (a > 0, \; b > 0).$$

The equality holds if and only if $a = b$. Note that because both a and b are positive, their arithmetic and geometric means are both positive, hence taking the square roots at the last step of the proof above is justified. Proofs of the other two inequalities are left for the reader. (Remember, the trick in teaching mathematics is: I do the easy part, and you do the hard part.)

Exercise. Can you generalize the inequalities above?

The following nice proof of the Pythagorean theorem is said to be given by an American middle school student. Unfortunately, I do not know his name, school, or when. First we need the following very useful theorem.

Theorem. Given four points A, B, C, D, let P be the intersection of lines AB and CD. Then points A, B, C, D are cocyclic if and only if

$$\overline{PA} \cdot \overline{PB} = \overline{PC} \cdot \overline{PD}.$$

(Vectors $\overrightarrow{PA}$ and $\overrightarrow{PB}$ are of the same (opposite) direction if and only if vectors $\overrightarrow{PC}$ and $\overrightarrow{PD}$ are of the same (opposite) direction.)

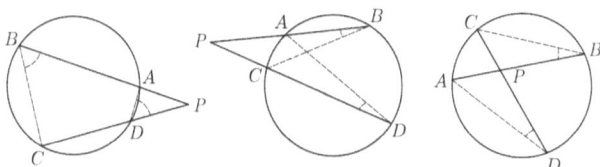

Figure PYG6

Proof. Proof is easy.

A, B, C, D, are cocyclic
$\iff \angle PBC = \angle PDA$
$\iff \triangle PBC \sim \triangle PDA$ (because we always have $\angle BPC = \angle DPA$)
$\iff \dfrac{\overline{PB}}{\overline{PC}} = \dfrac{\overline{PD}}{\overline{PA}}$
$\iff \overline{PA} \cdot \overline{PB} = \overline{PC} \cdot \overline{PD}$.

$\square$

In particular, if points C and D coincide, then we obtain the following.

Corollary. Suppose point P is on the extension of line segment AB (hence P is not between points A and B), and C is a point not on line AB. Then PC is tangent to the circumcircle of $\triangle ABC$ if and only if

$$\overline{PC}^2 = \overline{PA} \cdot \overline{PB}.$$

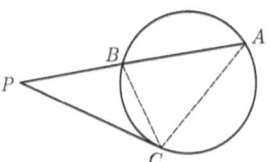

Figure PYG7

With the help of the corollary, the proof of the Pythagorean theorem is immediate. Given a right triangle ABC ($\angle C = 90°$), let D and E be the intersections of line AB and the circle centered at B with radius $\overline{BC}$. (Figure PYG8) Then AC is tangent to the circumcircle of $\triangle CDE$ (because AC is perpendicular to the radius BC), and so we have

$$\begin{aligned}\overline{AC}^2 &= \overline{AD} \cdot \overline{AE} = (\overline{AB} - \overline{BD}) \cdot (\overline{AB} + \overline{BE}) \\ &= (\overline{AB} - \overline{BC}) \cdot (\overline{AB} + \overline{BC}) = \overline{AB}^2 - \overline{BC}^2.\end{aligned}$$

Therefore, $\overline{AC}^2 + \overline{BC}^2 = \overline{AB}^2$.

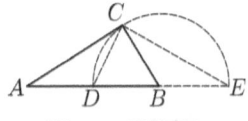

Figure PYG8

The Pythagorean Theorem

3.2 A Puzzle

Puzzle.

(a) Given two arbitrary squares, how do you cut them into (at most) five pieces so that the pieces can be rearranged to form a square? How many solutions can you find?

(b) Can you cut the big square into two pieces and the small one into three pieces (so that the pieces can be rearranged to form a square)?

(c) Is there a case we can achieve the purpose by cutting into a fewer (than five) pieces?

In case you have a difficulty in solving this puzzle, place one of the squares in Figure PYG on top of the other. Or, try the following exercises first.

Exercises.

1. Prove the Pythagorean theorem using Figure PYG9. (One proof for each figure.)

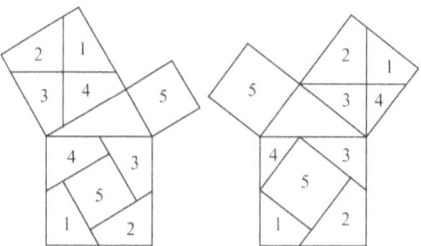

Figure PYG9

2. How about using Figure PYG10?

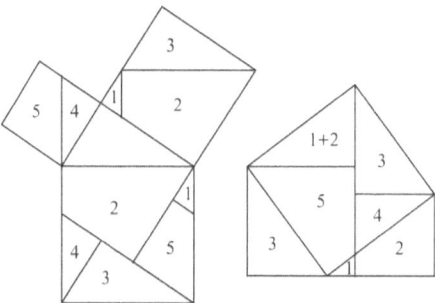

Figure PYG10

3. (a) Observe that

$$\begin{aligned}
3^2 + 4^2 &= 5^2 \\
20^2 + 21^2 &= 29^2 \\
119^2 + 120^2 &= 169^2 \\
696^2 + 697^2 &= 985^2 \\
4059^2 + 4060^2 &= 5741^2
\end{aligned}$$

What's next?

(b) Is there a positive integer solution to

$$m^2 + (m+1)^2 = n^4$$

other than $m = 119$, $n = 13$?

3.3 Pythagorean Triples

In the last chapter, we presented a formula for producing Pythagorean triples (page 8), but it did not give us all such triples. (Why?) So we now derive a formula that produces all the Pythagorean triples.

Let D, E, F be the points where the incircle $I(r)$ of right triangle ABC ($\angle C = 90°$) touches the three sides. (Figure PYG11) Then $IDCE$ is a square with side length r. Set

$$\overline{BF} = \overline{BD} = s, \quad \overline{AE} = \overline{AF} = t.$$

Before we proceed, we remark that $2r = a + b - c$ (Why?), and so if side lengths a, b, c are all integers, then inradius r must also be an integer (because the number of odd integers among a, b, c must be either 0 or 2), implying that both s and t are also integers.

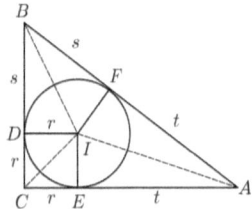

Figure PYG11

We have

$$[ABC] = \frac{1}{2}\overline{BC} \cdot \overline{CA} = \frac{1}{2}(r+s)(r+t).$$

The Pythagorean Theorem

On the other hand,
$$\begin{aligned}[ABC] &= [IBC] + [ICA] + [IAB]\\ &= \frac{r}{2}\{(s+r) + (r+t) + (t+s)\} = r(r+s+t).\end{aligned}$$

Therefore,
$$\frac{1}{2}(r+s)(r+t) = r(r+s+t).$$

Solving for t, we obtain
$$t = \frac{r(r+s)}{s-r}.$$

Note that $s > r$. (Why?) Thus
$$\begin{aligned}\overline{BC} &= a &&= s + r,\\ \overline{CA} &= b &&= r + t = r + \frac{r(r+s)}{s-r} = \frac{2rs}{s-r},\\ \overline{AB} &= c &&= s + t = s + \frac{r(r+s)}{s-r} = \frac{s^2+r^2}{s-r}.\end{aligned}$$

Therefore,
$$a : b : c = (s+r) : \left(\frac{2rs}{s-r}\right) : \left(\frac{s^2+r^2}{s-r}\right) = \left(s^2 - r^2\right) : (2rs) : \left(s^2 + r^2\right).$$

Note that we are not assuming that a, b, c are integers, hence our derivation shows that the lengths of the three sides of any right triangle must satisfy this ratio for some r and s; i.e., for some constant $k\ (= 1/(s-r))$,
$$a = k\left(s^2 - r^2\right), \quad b = k(2rs), \quad c = k\left(s^2 + r^2\right).$$

It follows that
$$a^2 + b^2 = k^2\left\{\left(s^2 - r^2\right)^2 + (2rs)^2\right\} = \left\{k\left(s^2+r^2\right)\right\}^2 = c^2.$$

We obtained yet another proof of the Pythagorean theorem.

Now, for a Pythagorean triple, if both r and s are integers of the same parity (i.e., both even or both odd), then the lengths of all three sides are even numbers, hence not relatively prime. Thus to obtain a "primitive" Pythagorean triple, r and s must be of different parities.

Remark. Recall the trigonometric identities
$$\begin{aligned}\sin\theta &= 2\sin\frac{\theta}{2}\cdot\cos\frac{\theta}{2} = \frac{2\sin\frac{\theta}{2}\cdot\cos\frac{\theta}{2}}{\cos^2\frac{\theta}{2}+\sin^2\frac{\theta}{2}} = \frac{2\tan\frac{\theta}{2}}{1+\tan^2\frac{\theta}{2}},\\ \cos\theta &= \cos^2\frac{\theta}{2} - \sin^2\frac{\theta}{2} = \frac{\cos^2\frac{\theta}{2}-\sin^2\frac{\theta}{2}}{\cos^2\frac{\theta}{2}+\sin^2\frac{\theta}{2}} = \frac{1-\tan^2\frac{\theta}{2}}{1+\tan^2\frac{\theta}{2}}.\end{aligned}$$

Choosing $\theta = \angle ABC$ (Figure PYG11), we have

$$\tan\frac{\theta}{2} = \tan(\angle IBD) = \frac{\overline{ID}}{\overline{BD}} = \frac{r}{s}.$$

Substituting this expression into the identities above, we obtain

$$\sin\theta = \frac{2\left(\frac{r}{s}\right)}{1+\left(\frac{r}{s}\right)^2} = \frac{2rs}{s^2+r^2}, \qquad \cos\theta = \frac{1-\left(\frac{r}{s}\right)^2}{1+\left(\frac{r}{s}\right)^2} = \frac{s^2-r^2}{s^2+r^2}.$$

This is the idea behind our derivation above.

Exercise. Suppose a, b, c are integers satisfying $a^2+b^2=c^2$. Show that

(a) at least one of a and b is divisible by 3;

(b) at least one of a and b is divisible by 4;

(c) at least one of a, b, and c is divisible by 5.

3.4 Generalizations of the Pythagorean Theorem

We now present Euclid's proof of the Pythagorean theorem, and show that it can easily be extended to obtain the law of cosines.

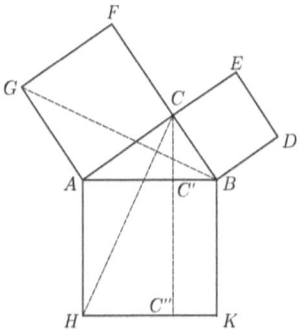

Figure PLC

Let ABC be a right triangle where $\angle ACB = 90°$. Draw squares $BDEC$, $CFGA$ and $AHKB$ outwardly on respective sides BC, CA and AB, of $\triangle ABC$. (Figure PLC) Then it is easy to see that rotating $\triangle AHC$ by $90°$, we obtain $\triangle ABG$; viz.,

$$AH \perp AB, \qquad AC \perp AG,$$

The Pythagorean Theorem

and so

$$\angle HAC = \angle HAB + \angle BAC = 90° + \angle BAC$$
$$= \angle BAC + \angle CAG = \angle BAG.$$

Furthermore,

$$\overline{AH} = \overline{AB}, \quad \overline{AC} = \overline{AG}.$$

Therefore,

$$\triangle AHC \cong \triangle ABG \quad (SAS).$$

Because $BC \parallel AG$, square $CFGA$ and $\triangle ABG$ have the same height AC on their common side AG, we have

$$[CFGA] = 2[ABG].$$

Similarly,

$$[AHC''C'] = 2[AHC],$$

where C' and C'' are the feet of the perpendicular from C to AB and HK, respectively. But $\triangle ABG \cong \triangle AHC$, and so

$$[CFGA] = 2[ABG] = 2[AHC] = [AHC''C'].$$

Similarly, rotating $\triangle BDA$ by $90°$, we obtain $\triangle BCK$, and so

$$[BDEC] = [BC'C''K].$$

It follows that

$$[BDEC] + [CFGA] = [BC'C''K] + [AHC''C'] = [AHKB];$$

But

$$[BDEC] = \overline{BC}^2 = a^2, \quad [CFGA] = \overline{CA}^2 = b^2, \quad [AHKB] = \overline{AB}^2 = c^2,$$

and so we obtain

$$a^2 + b^2 = c^2.$$

Now let us investigate what happens in a general case. We keep all the notations as in the proof above, and let A' and A'' be the feet of the perpendicular from A to BC and DE, respectively. Similarly, for B' and B''. (Figure PLC1)

Then, as before, rotating $\triangle AHC$ by $90°$, we obtain $\triangle ABG$, but the height of $\triangle ABG$ on side AG is AB', and so

$$[AHC''C'] = 2[AHC] = 2[ABG] = [AB'B''G].$$

Similarly,

$$[BC'C''K] = [BDA''A'].$$

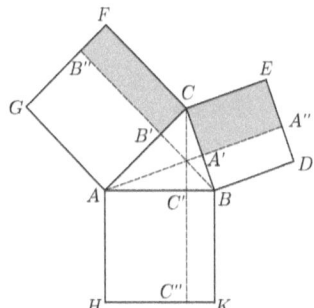

Figure PLC1

Thus

$$\begin{aligned}
a^2 + b^2 &= [BDEC] + [CFGA] \\
&= ([BDA''A'] + [CA'A''E]) + ([CFB''B'] + [AB'B''G]) \\
&= ([BDA''A'] + [AB'B''G]) + ([CA'A''E] + [CFB''B']) \\
&= ([BC'C''K] + [AHC''C']) + ([CA'A''E] + [CFB''B']) \\
&= [AHKB] + \overline{CA'} \cdot \overline{CE} + \overline{CF} \cdot \overline{CB'} \\
&= c^2 + (\overline{CA} \cdot \cos \angle C) \cdot a + b \cdot (\overline{BC} \cdot \cos \angle C) \\
&= c^2 + 2ab \cdot \cos \angle C,
\end{aligned}$$

which is the law of cosines.

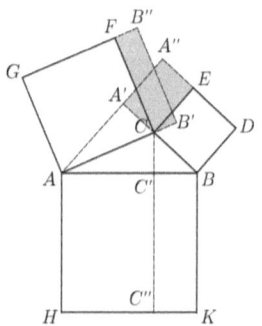

Figure PLC2

Note carefully that our proof is valid regardless of whether $\angle C$ is acute or obtuse without any change provided we interpret the area of a rectangle to be positive whenever the rectangle is labelled counterclockwise, but negative if labelled clockwise.

Finally, we modify Euclid's proof to obtain yet another generalization of the Pythagorean theorem. Again we preserve the notations used above, and let C_0

The Pythagorean Theorem

be the intersection of DE and FG; A_0 the intersection of AH and FG; B_0 the intersection of BK and DE. (Figure PLC3)

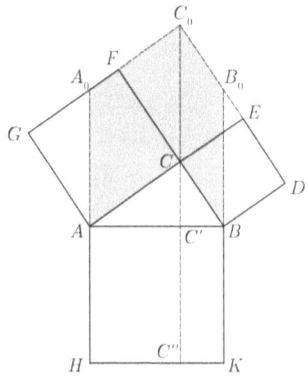

Figure PLC3

Then CEC_0F is a rectangle, and
$$\triangle C_0CE \cong \triangle ABC \quad (SAS),$$
and so C_0, C, C', C'' are collinear. Thus BB_0C_0C and CC_0A_0A are parallelograms, and
$$[BDEC] = [BB_0C_0C], \quad [CFGA] = [CC_0A_0A].$$
Furthermore, $\overline{C_0C} = \overline{AB} = \overline{C'C''}$. Therefore,
$$[BB_0C_0C] = [BC'C''K], \quad [CC_0A_0A] = [AHC''C'].$$
It follows that
$$\begin{aligned} a^2 + b^2 &= [BDEC] + [CFGA] = [BB_0C_0C] + [CC_0A_0A] \\ &= [BC'C''K] + [AHC''C'] = [AHKB] = c^2. \end{aligned}$$

Remark. Observe that $\triangle A_0B_0C_0 \cong \triangle ABC$. This observation plays a role in the proof of the last theorem in this chapter.

Imitating this proof, we obtain the following, which is also a generalization of the Pythagorean theorem.

Theorem 2. Draw arbitrary parallelograms $BDEC$ and $CFGA$ outwardly on sides BC and CA of an arbitrary triangle ABC. (Figure PLC4) Let C_0 be the intersection of DE and FG, and draw parallelogram $AHKB$ such that
$$AH \parallel C_0C \parallel BK, \quad \overline{AH} = \overline{C_0C} = \overline{BK}.$$
Then
$$[BDEC] + [CFGA] = [AHKB].$$

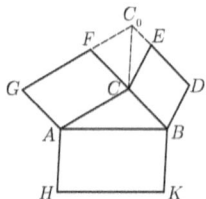

Figure PLC4

Theorem 3. Suppose PQ is a line segment in the plane of $\triangle ABC$. Let the perpendicular projections of point P on sides BC, CA, AB be P_1, P_2, P_3, respectively; and that of point Q be Q_1, Q_2, Q_3. (Figure PLC5) Then
$$\overline{BC} \cdot \overline{P_1Q_1} + \overline{CA} \cdot \overline{P_2Q_2} + \overline{AB} \cdot \overline{P_3Q_3} = 0,$$
where $\overline{BC} \cdot \overline{P_1Q_1} > 0$ if $\overrightarrow{BC}$ and $\overrightarrow{P_1Q_1}$ have the same direction, but $\overline{BC} \cdot \overline{P_1Q_1} < 0$ if they have the opposite directions. Similarly, for $\overline{CA} \cdot \overline{P_2Q_2}$ and $\overline{AB} \cdot \overline{P_3Q_3}$.

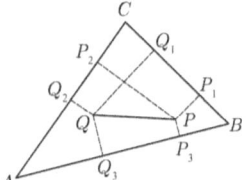

Figure PLC5

We present only a hint and leave the details for the reader. (Recall the remark on page 23.) Let A_0, B_0, C_0 be points (Figure PLC6) such that
$$\overline{AA_0} = \overline{BB_0} = \overline{CC_0} = \overline{PQ} \quad \text{and} \quad \overrightarrow{AA_0} \parallel \overrightarrow{BB_0} \parallel \overrightarrow{CC_0} \perp \overrightarrow{PQ}.$$
Then consider the signed areas of parallelograms AA_0B_0B, BB_0C_0C, CC_0A_0A.

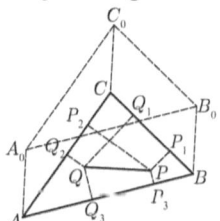

Figure PLC6

Note that if ABC is a right triangle, and PQ coincides with the hypotenuse, then the theorem reduces to the Pythagorean theorem.

Chapter 4
Japanese Temple Mathematics

From 17th century to early 20th century, people of various walks in Japan offered their mathematical discoveries on wooden tablets to shrines and temples. Most of the time, it was done for pure enjoyment or to challenge others (but certainly not for college entrance or job promotion). Majority of them were in geometry. To me that is a strong indication that geometry had (and still has) popular appeal to the general public. Thus I find it very tragic that geometry is being de-emphasized in school mathematics nowadays.

We now present a few samples of the Japanese temple mathematics. Readers may try to find solutions without using trigonometry or coordinates as such tools were not available in Japan in those days.

4.1 Problems

1. (1856) In Figure JTM1, each of the circles has radius 1 and touches its neighbors, but no two of them overlap each other. Find the total area inside the circles but outside the polygon whose vertices are at the centers of the circles (in terms of the number of the circles).

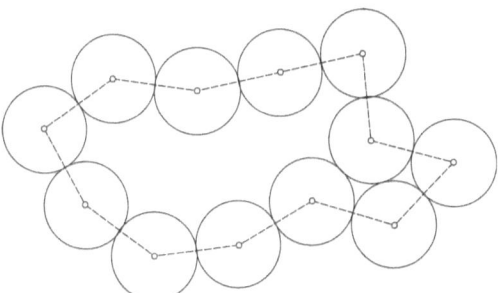

Figure JTM1

2. A right triangle is inscribed in a semi-circle. (Figure JTM2) If the radius of the maximal circle inscribed in the region bounded by the semi-circle and a leg of the right triangle is equal to that of the incircle of the right triangle, find the ratio between the radii of the semi-circle and the incircle.

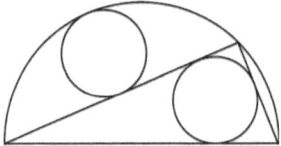

Figure JTM2

3. (1743) 7 melons cost 37 mon, 10 eggplants cost 7 mon, 9 peaches cost 8 mon. How can one buy 960 of them with 960 mon? ("Mon" was an old Japanese currency.)

4. (1826) In Figure JTM4, squares $ABCD$, $BEFG$, $CGHJ$, $DJKL$ have side lengths a, b, c, d, respectively. Show that E, A, L are collinear if and only if $2a = c$.

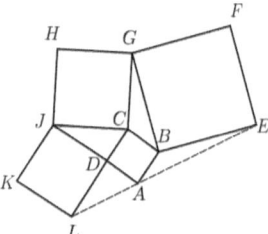

Figure JTM4

5. (1797) A square is divided into a triangle and a trapezoid. If two equal circles are inscribed in the trapezoid as in Figure JTM5, find the radius of the circles in terms of the side length of the square.

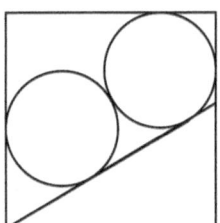

Figure JTM5

6. Suppose ABC is an isosceles triangle ($\overline{AB} = \overline{BC}$) with two vertices B and C on circle O. If four equal circles are inscribed as in Figure JTM6, find the ratio

between the radii of the big circle and the four equal circles.

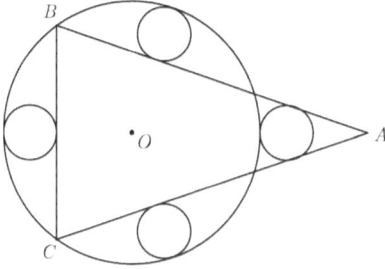

Figure JTM6

4.2 Solutions

1. The solution is immediate if we remember that the algebraic sum of the exterior angles of a simple polygon is 360°, provided we consider the exterior angle to be positive if the interior angle is less than 180°, but negative if the interior angle is more than 180°.

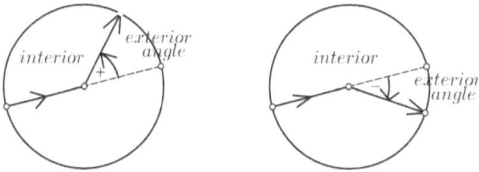

Figure JTM1'

Because 360° corresponds to a full circle, the area inside both the polygon and the union of the circles is half of the area of the union of the circles minus one circle; i.e., $(\frac{n}{2} - 1) \cdot \pi$ (assuming there are n circles, $n \geq 3$); while the area inside the union of the circles but outside the polygon is half of the union of the circles plus one circle; i.e., $(\frac{n}{2} + 1) \cdot \pi$.

Note that the difference of the two areas is always equal to the area of two circles regardless of the number of the circles.

2. Let right triangle ABC be inscribed in semi-circle $O(R)$; incircle $I(r)$ of $\triangle ABC$ touches three sides BC, CA, AB at D, E, F, respectively; the perpendicular bisector of CA intersects CA and $\overset{\frown}{CA}$ at G and J, respectively; and finally, the lengths of three sides BC, CA, AB be a, b, c, respectively. (Figure JTM2')
Then
$$R = \overline{OJ} = \overline{OG} + \overline{GJ} = \frac{a}{2} + 2r, \quad a = 2(R - 2r).$$

On the other hand,

$$\begin{aligned}
2R &= c = \overline{AB} = \overline{AF} + \overline{BF} = \overline{AE} + \overline{BD} \\
&= (\overline{AC} - \overline{CE}) + (\overline{BC} - \overline{CD}) = (b-r) + (a-r) \\
&= a+b-2r = 2(R-2r) + b - 2r, \\
b &= 6r.
\end{aligned}$$

By the Pythagorean theorem: $a^2 + b^2 = c^2$,

$$4(R-2r)^2 + (6r)^2 = (2R)^2, \quad 13r = 4R.$$

The solution above is due to Mr. Dean Ballard of Lakeside School, Seattle.

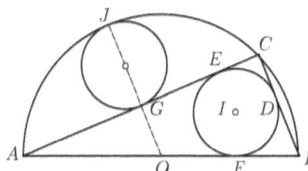

Figure JTM2'

Or, we may replace the Pythagorean theorem at the last step of the solution above by considering the areas. Because

$$\begin{aligned}
[ABC] &= [IBC] + [ICA] + [IAB] \\
\tfrac{1}{2}ab &= \tfrac{r}{2}(a+b+c). \\
2(R-2r) \cdot (6r) &= r\{2(R-2r) + 6r + 2R\}, \\
4R &= 13r.
\end{aligned}$$

Note that $\overline{BC} : \overline{CA} : \overline{AB} = 5 : 12 : 13$.

3. Let x, y and z be the numbers of melons, eggplants and peaches, respectively. Then we have a system of simultaneous linear equations

$$x + y + z = 960, \quad \frac{37}{7}x + \frac{7}{10}y + \frac{8}{9}z = 960.$$

We claim: 7 divides x; 10 divides y; and 9 divides z. For, multiplying both sides of the second equation by $7 \cdot 10 \cdot 9$, we obtain

$$(37 \cdot 10 \cdot 9)x + (7^2 \cdot 9)y + (8 \cdot 7 \cdot 10)z = 960 \cdot 7 \cdot 10 \cdot 9,$$

where all the terms except the first are divisible by 7. Therefore, the first term must also be divisible by 7. But 7 and $37 \cdot 10 \cdot 9$ are relatively prime, hence 7 must divide

Japanese Temple Mathematics

x. Similarly for the other two terms. Thus we may set $x = 7a$, $y = 10b$, $z = 9c$. Then the system of simultaneous linear equations becomes

$$7a + 10b + 9c = 960, \quad 37a + 7b + 8c = 960.$$

From the difference of these two equations, we obtain $c = 30a - 3b$. Substituting this expression into one of the simultaneous linear equations above, we obtain

$$7a + 10b + 9(30a - 3b) = 960, \quad 277a - 17b = 960.$$
$$17b = 277a - 960 = 17(16a - 56) + (5a - 8).$$

Thus $5a - 8$ must be divisible by 17. And we may write

$$5a - 8 = 17t, \quad 5a = 17t + 8 = 5(3t + 1) + (2t + 3),$$

implying that $2t + 3$ is divisible by 5. But then t must be of the form $t = 5s + 1$ for some s (because $3(2t + 3) = 5(t + 2) + (t - 1)$, $t - 1$ must be divisible by 5. Set $t - 1 = 5s$). Substituting this expression into $5a - 8 = 17t$, we obtain

$$5a = 17(5s + 1) + 8 = 5(17s + 5), \quad a = 17s + 5.$$

From $37a + 7b + 8c = 960$, we obtain $a < \frac{960}{37} = 25.9\cdots$, hence $s = 0$, or 1; i.e., $a = 5$ or 22. Substituting $a = 5$ into $277a - 17b = 960$, we obtain

$$b = \frac{277 \cdot 5 - 960}{17} = \frac{1385 - 960}{17} = 25,$$
$$c = 30a - 3b = 30 \cdot 5 - 3 \cdot 25 = 75.$$
$$x = 7a = 35, \quad y = 10b = 250, \quad z = 9c = 675.$$

Now if $a = 22$, then

$$b = \frac{277 \cdot 22 - 960}{17} = \frac{6094 - 960}{17} = \frac{5134}{17} = 302,$$

giving $y = 10b = 3020 > 960$. So this case cannot happen. Thus the only answer is 35 melons, 250 eggplants, and 675 peaches.

Remark. All solutions of the original system of simultaneous linear equations can be expressed as

$$x = 7(17s + 5), \quad y = 10(277s + 25), \quad z = 9(-321s + 75).$$

For x, y, z to be all positive, $s = 0$ is the only choice.

4. Let M and N be the midpoints of CG and CJ, respectively. (Figure JTM4a)

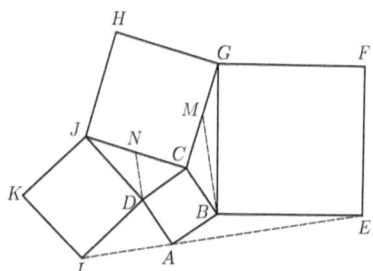

Figure JTM4a

First, suppose $c = 2a$; i.e., $\overline{CD} = \overline{CB} = \overline{CM} = \overline{CN}$. Then

$$\triangle CDB \cong \triangle CMN \ (SAS), \text{ and so } \overline{DB} = \overline{MN}.$$

But points D, B, M, N are cocyclic (centered at C), and hence $BM \parallel DN$. (Figure JTM4b/Left) Now from the lemma below, we have

$$AE \perp BM \parallel DN \perp AL,$$

implying that A, E, L are collinear.

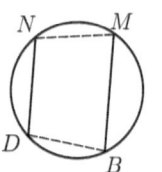 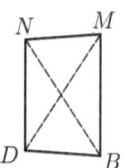

Figure JTM4b

Conversely, suppose A, E, L are collinear. (Figure JTB4a) We want to show that $\overline{CD} = \overline{CM}$ (where M and N are the midpoints of CG and CJ, respectively). Rotating $\triangle CDM$ by $90°$ around C, we obtain $\triangle CBN$, and so

$$\triangle CDM \cong \triangle CBN \ (SAS), \text{ hence } \overline{DM} = \overline{BN}.$$

But $BM \parallel DN$ (because $BM \perp AE \parallel AL \perp DN$, by the lemma below), implying that $BMND$ is an isosceles trapezoid; i.e., $\overline{DB} = \overline{MN}$. (Figure JTM4b/Right) Thus $\overline{CD} = \overline{CM}$ (because $\triangle CDB \sim \triangle CMN$).

It remains to prove the following.

> **Lemma.** Let $BDEC$ and $CFGA$ be squares on sides BC and CA of $\triangle ABC$, respectively, and M a point on EF. (Figure JTM4c) Then
>
> $$M \text{ is the midpoint of } EF \quad \Longleftrightarrow \quad CM \perp AB.$$

Japanese Temple Mathematics 31

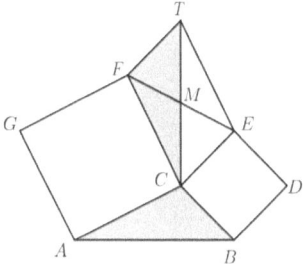

Figure JTM4c

Proof. Suppose M is the midpoint of EF. Extend CM to T such that M is also the midpoint of CT. Then clearly, $CETF$ is a parallelogram. In $\triangle ABC$ and $\triangle CTF$, two pairs of sides; i.e., $\{CA, FC\}$ and $\{BC, TF\}$, are equal and perpendicular, hence

$$\triangle ABC \cong \triangle CTF \quad (SAS),$$

which, in turn, implies that the third pair, $\{AB, CT\}$, to be perpendicular; i.e., $CM \perp AB$. □

Because the converse was not needed in our proof above, we leave its proof as an exercise for the reader. (*Hint.* This time let T be the intersection of CM and the line passing through F and perpendicular to BC; i.e., parallel to CE. Then we still have $\triangle ABC \cong \triangle CTF$.) Actually, the midpoint of EF is unique, so is the line passing through C perpendicular to AB, and so the converse is trivial. (See my book, *Honsberger Revisited*, National Taiwan University Press, 2012; pp. 67-71.)

5. In Figure JTM5', let E be the third vertex of the triangle; O_1, O_2 the centers of the two equal circles with radius r; F the foot of the perpendicular from O_2 to side DA; G the foot of the perpendicular from O_1 to O_2F; and $\overline{BC} = a$, $\overline{CE} = b$, $\overline{BE} = c$.

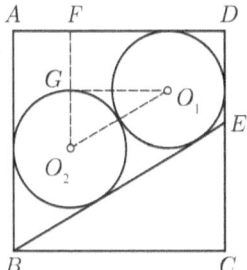

Figure 4.1: JTM5'

Clearly, O_1 is on diagonal BD, and so

$$[BED] = [O_1BE] + [O_1ED],$$
$$2[BED] = \overline{BC} \cdot \overline{ED} = a(a-b), \quad 2[O_1BE] = rc, \quad 2[O_1ED] = r(a-b).$$

Substituting these expressions into the formula of $[BED]$, we obtain

$$a(a-b) = r(c+a-b), \quad r = \frac{a(a-b)}{c+a-b}.$$

On the other hand,

$$[ABED] = [O_2AB] + [O_2BE] + [O_2ED] + [O_2DA],$$
$$2[ABED] = \overline{AD} \cdot (\overline{AB} + \overline{DE}) = a(2a-b);$$
$$2[O_2AB] = ra, \quad 2[O_2BE] = rc, \quad 2[O_2ED] = \overline{ED} \cdot \overline{DF} = (a-b)(a-r).$$

Obviously, $\triangle O_1GO_2 \sim \triangle BCE$, and so

$$\frac{\overline{O_2G}}{\overline{O_1O_2}} = \frac{\overline{EC}}{\overline{BE}}, \quad \overline{O_2G} = \frac{\overline{EC}}{\overline{BE}} \cdot \overline{O_1O_2} = \frac{b}{c}(2r).$$

$$2[O_2DA] = \overline{DA} \cdot \overline{O_2F} = a \cdot (\overline{O_2G} + \overline{GF}) = a\left(\frac{2b}{c} + 1\right)r = \frac{a(2b+c)}{c}r.$$

Substituting these expressions into the formula of $[ABED]$, we obtain

$$a(2a-b) = r(a+c) + (a-b)(a-r) + \frac{a(2b+c)}{c}r.$$

Solving for r from this equality, we obtain

$$r = \frac{ca^2}{c^2 + (a+b)c + 2ab} = \frac{ca^2}{(a+b)(a+b+c)},$$

where we used $c^2 = a^2 + b^2$. Equating the two expressions of r, we obtain

$$\frac{a-b}{c+a-b} = \frac{ca}{(a+b)(a+b+c)}.$$

Rewriting this equality (and using $c^2 = a^2 + b^2$ again), we obtain

$$(a-b)c = -a^2 + 2ab + b^2.$$

Squaring both sides (and using $c^2 = a^2 + b^2$ again), we obtain

$$(a-b)^2 (a^2+b^2) = (-a^2 + 2ab + b^2)^2.$$

This equality reduces to

$$a^2 = 3b^2, \quad a = \sqrt{3}\, b, \quad c = 2b.$$

Therefore,

$$r = \frac{a(a-b)}{c+a-b} = a \cdot \frac{(\sqrt{3}-1)b}{(2+\sqrt{3}-1)b} = a \cdot \frac{\sqrt{3}-1}{\sqrt{3}+1} = \frac{(\sqrt{3}-1)^2}{2} a = \left(2-\sqrt{3}\right) a.$$

6. Let R and r be the radii of the big circle and the four equal circles, respectively; D the feet of the perpendicular from center O to AB; E the center of one of the four equal circles closest to vertex A; F the midpoint of side BC; and K the intersection of OD and the line passing through E parallel to AB. (Figure JTM6')

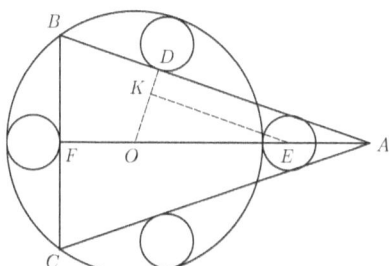

Figure JTM6'

Clearly, $\triangle OAD \sim \triangle OEK$, and so

$$\frac{\overline{OA}}{\overline{OD}} = \frac{\overline{OE}}{\overline{OK}}, \quad \overline{OA} = \frac{\overline{OE}}{\overline{OK}} \cdot \overline{OD} = \frac{R+r}{R-3r} \cdot (R-2r),$$

$$\overline{AF} = \overline{OA} + \overline{OF} = \frac{(R+r)(R-2r)}{R-3r} + (R-2r)$$

$$= \frac{(R-2r)\{(R+r)+(R-3r)\}}{R-3r} = \frac{2(R-2r)(R-r)}{R-3r}.$$

By the Pythagorean theorem,

$$\overline{BF}^2 = \overline{OB}^2 - \overline{OF}^2 = R^2 - (R-2r)^2 = 4r(R-r).$$

We also have $\triangle ABF \sim \triangle EOK$, hence

$$\frac{\overline{AB}}{\overline{BF}} = \frac{\overline{EO}}{\overline{OK}}, \quad \overline{AB} = \frac{\overline{EO}}{\overline{OK}} \cdot \overline{BF} = \frac{R+r}{R-3r} \cdot \sqrt{4r(R-r)}.$$

Applying the Pythagorean theorem to $\triangle ABF$, we have

$$\overline{AB}^2 = \overline{BF}^2 + \overline{AF}^2,$$

$$\left(\frac{R+r}{R-3r}\right)^2 \cdot \{4r(R-r)\} = 4r(R-r) + \left\{\frac{2(R-2r)(R-r)}{R-3r}\right\}^2.$$

Clearing the denominator and collecting the like terms, this equality reduces to

$$R^3 - 5R^2 r + 4r^3 = 0, \quad (R-r)\left(R^2 - 4Rr - 4r^2\right) = 0.$$

Because $R \neq r$, and $R > 0$, we conclude that

$$R = 2\left(1 + \sqrt{2}\right) r.$$

Exercises

1. Two equal circles are inscribed in a right triangle as in Figure JTM1e. Express the radius of the circles in terms of the side lengths of the right triangle.

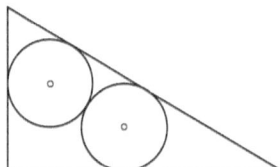

Figure JTM1e

2. An equilateral triangle is divided into three congruent triangles and an equilateral triangle. (Figure JTM2e) Assuming these four incircles are all equal, find the (common) radius in terms of the side length of the outside equilateral triangle.

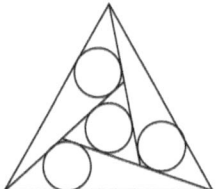

Figure JTM2e

3. Suppose point P is inside circle O. (Figure JTM3e) Two circles (with radii r_1 and r_2) are tangent to each other externally at P and each of them is tangent to circle O. Show that
$$\frac{1}{r_1} + \frac{1}{r_2} \text{ is a constant.}$$
What if point P is outside circle O?

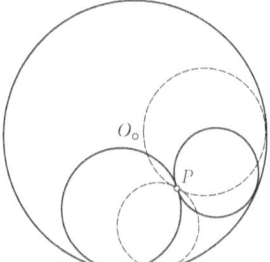

Figure JTM3e

4. (a) Suppose ABC is an equilateral triangle with vertices B, C on line XY, and line UV passes through vertex A. (Figure JTM4e/Left) Furthermore, circle $O_1(r_1)$ touches lines UV, XY, and AB; and circle $O_2(r_2)$ touches lines UV, XY, and AC. Show that, as line UV rotates about vertex A, $r_1 + r_2$ remains constant.

   (b) Actually, all we need is for $\triangle ABC$ to be isosceles; i.e., $\overline{AB} = \overline{AC}$. In other words, $r_1 + r_2$ depends neither on the length of BC, nor on line UV. (Figure JTM4e/Right) Prove this generalization.

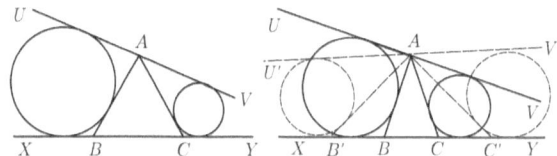

Figure JTM4e

5. A rooster costs 5 sen, a hen costs 3 sen, and three little chickens cost 1 sen. How do you buy 100 of them with 1 yen (= 100 sen)? Is the answer unique?

The following problem, posed and solved by ŌMURA Kingo Kazuhide in 1841, is communicated to me by Professor TSUCHIKURA Tamotsu of Tohoku University, Sendai, Japan.

> Given three spheres that are tangent to plane Π (on the same side of Π), and also pairwise tangent to each other externally, find the largest regular tetrahedron with one of the faces on plane Π, and each of the remaining faces is tangent to one of the given spheres such that different faces are tangent to different spheres.

We leave this problem as well as its generalization (that the three spheres are not necessarily pairwise tangent to each other) as an exercise. However, as a hint, we present a solution to the following problem on a plane.

Given three circles on a plane, find the largest equilateral triangle whose three sides are tangent to the three given circles such that different sides are tangent to different circles. (Figure JTM7)

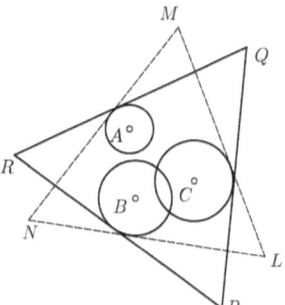

Figure JTM7

Solution. Let three given circles be $A(r_1)$, $B(r_2)$, $C(r_3)$, and $\triangle PQR$ the desired largest equilateral triangle. To compare the sizes of "competing" equilateral triangles, we reduce the altitude of each of these equilateral triangles, say LMN, by a fixed length; i.e., we do parallel translation of each side of $\triangle LMN$ such that the new side passes through the center of (instead of tangent to) the corresponding circle. (JTM8/Left) Clearly, the altitude of the (new) equilateral triangle, denoted by $L'M'N'$, is shorter than that of the (original) equilateral triangle LMN by the sum of the three radii, $r_1 + r_2 + r_3$.

So now we have only to solve the particular case that $r_1 = r_2 = r_3 = 0$:

Find the largest equilateral triangle each of its three sides passes through one of the three given points such that different sides pass through different points.

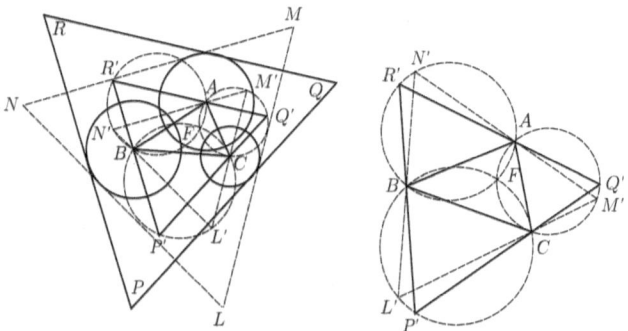

Figure JTM8

Japanese Temple Mathematics

Suppose equilateral triangle $L'M'N'$ satisfies the condition that each of its sides passes through one of the points, A, B, C (different sides pass through different points); and $\triangle P'Q'R'$ is the largest among such equilateral triangles. (Figure JTM8) Then $\angle BL'C = \angle BP'C\,(=\frac{\pi}{3})$, and so points L', P', B, C, are cocyclic. Similarly, points M', Q', C, A, are cocyclic; so are points N', R', A, B. Furthermore, these three circles; i.e., the circumcircles of quadrangles, $L'P'BC$, $M'Q'CA$, and $N'R'AB$, meet at a point (Why?), say at F.

In $\triangle FL'M'$ and $\triangle FP'Q'$, because $\angle FL'M' = \angle FL'C = \angle FP'C = \angle FP'Q'$; and similarly, $\angle FM'L' = \angle FQ'P'$, we have

$$\triangle FL'M' \sim \triangle FP'Q'.$$

And so instead of comparing the lengths of $L'M'$ and $P'Q'$, we may do that for, say FL' and FP'. But these are the chords of the circumcircle of quadrangle $L'P'BC$, and hence the largest case happens if and only if FP' is a diameter (of the circumcircle).

Note carefully that if FP' is a diameter of the circumcircle of quadrangle $L'P'BC$, then FQ' must be a diameter of the circumcircle of quadrangle $M'Q'CA$. (Why?) Similarly, FR' must be a diameter of the circumcircle of quadrangle $N'R'AB$. Furthermore, the circumcircles of these three quadrangles are precisely the circumcircles of the equilateral triangles constructed externally on the three sides of $\triangle ABC$. Thus we obtain an algorithm to construct the required largest equilateral triangle:

1. Given circles $A(r_1)$, $B(r_2)$, $C(r_3)$, construct circumcircles of the three equilateral triangles that are drawn externally on the sides of $\triangle ABC$.

2. These three circumcircles share a point; call it F. Draw the diameter with F as one end to each of the circles. Then the three other ends give the vertices of equilateral triangle $P'Q'R'$.

3. Finally, expand $\triangle P'Q'R'$ by parallel translation of each of the sides such that each side of the equilateral triangle PQR is tangent to the corresponding circle among the three given ones (different sides are tangent to different circles).

We now consider the problem of calculating the side length of equilateral triangle $P'Q'R'$ (where FP' is a diameter of the circumcircle of $\triangle P'BC$). We know that the side length of an equilateral triangle can be expressed in terms of the distances from a point to the three vertices.[1]

Let $t = \overline{Q'R'} = \overline{R'P'} = \overline{P'Q'}$, $\theta = \angle FP'Q'$, and

$$p = \overline{FP'} = \frac{\overline{BC}}{\sin\frac{\pi}{3}} = \frac{2a}{\sqrt{3}}, \quad q = \overline{FQ'} = \frac{2b}{\sqrt{3}}, \quad r = \overline{FR'} = \frac{2c}{\sqrt{3}},$$

where we used the law of sines. Note that if point F is outside the equilateral triangle $P'Q'R'$, then θ could be negative or greater than $\frac{\pi}{3}$. (Figure JTM9)

---
[1] See my book, *Honsberger Revisited* (National Taiwan University Press, Taipei, 2012), Problem 2(b), page 136; Solution, page 170.

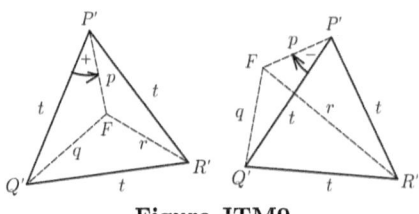

Figure JTM9

By the law of cosines, we have

$$q^2 = t^2 + p^2 - 2pt\cos\theta, \quad \cos\theta = \frac{t^2 + p^2 - q^2}{2pt}.$$

Similarly,

$$\begin{aligned}
r^2 &= t^2 + p^2 - 2pt\cos\left(\frac{\pi}{3} - \theta\right) \\
&= t^2 + p^2 - 2pt\left\{\frac{1}{2}\cos\theta + \frac{\sqrt{3}}{2}\sin\theta\right\} \\
&= t^2 + p^2 - 2pt\left\{\frac{1}{2}\cdot\frac{t^2 + p^2 - q^2}{2pt} + \frac{\sqrt{3}}{2}\sin\theta\right\} \\
&= \frac{1}{2}\left(t^2 + p^2 + q^2\right) - \sqrt{3}pt\sin\theta. \\
\sin\theta &= \frac{t^2 + p^2 + q^2 - 2r^2}{2\sqrt{3}pt}.
\end{aligned}$$

Substituting these expressions into the identity $\sin^2\theta + \cos^2\theta = 1$, we obtain a quartic polynomial in t,

$$\left(t^2 + p^2 + q^2 - 2r^2\right)^2 + 3\left(t^2 + p^2 - q^2\right)^2 - 12p^2t^2 = 0.$$

After a straightforward computation, this equation reduces to a quadratic polynomial in t^2,

$$\begin{aligned}
t^4 &- t^2\left(p^2 + q^2 + r^2\right) + \left(p^4 + q^4 + r^4 - q^2r^2 - r^2p^2 - p^2q^2\right) = 0. \\
t^2 &= \frac{1}{2}\left\{\left(p^2 + q^2 + r^2\right)\right. \\
&\qquad \left. \pm \sqrt{\left(p^2 + q^2 + r^2\right)^2 - 4\left(p^4 + q^4 + r^4 - q^2r^2 - r^2p^2 - p^2q^2\right)}\right\} \\
&= \frac{1}{2}\left\{\left(p^2 + q^2 + r^2\right) \pm \sqrt{3}\cdot\sqrt{2\left(q^2r^2 + r^2p^2 + p^2q^2\right) - \left(p^4 + q^4 + r^4\right)}\right\},
\end{aligned}$$

where the choice of plus or minus sign before the square root sign depends on whether point F is inside or outside $\triangle P'Q'R'$. We may now substitute $p = \frac{2a}{\sqrt{3}}$, $q = \frac{2b}{\sqrt{3}}$,

Japanese Temple Mathematics

$r = \frac{2c}{\sqrt{3}}$, to express t in terms of $\overline{BC}$, $\overline{CA}$, $\overline{AB}$.

$$t^2 = \frac{2}{3}\left\{(a^2+b^2+c^2) \pm \sqrt{3} \cdot \sqrt{2(b^2c^2+c^2a^2+a^2b^2)-(a^4+b^4+c^4)}\right\}.$$

Once we know the value of t, we can obtain the side length of the desired largest equilateral triangle PQR as

$$t + \csc\frac{\pi}{3}(r_1+r_2+r_3) = t + \frac{2}{\sqrt{3}}(r_1+r_2+r_3).$$

A careful reader might notice that our solution above assumes tacitly that none of the angles in $\triangle ABC$ is greater than $\frac{2\pi}{3}$. (Where did we use this assumption?) We leave this remaining case for the reader to explore. (I do the easy part, you do the hard part.) (Figure JTM10)

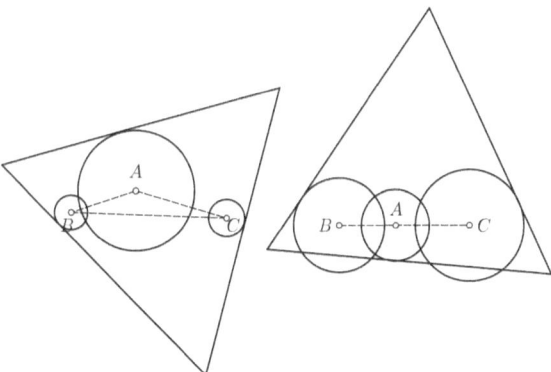

Figure JTM10

Chapter 5

Mind Reading Tricks

5.1 Trick 1

Here is a trick that is sure to be a hit. First we need to create 5 cards. Card A from Figure MRT, and all four cards from Figure MRT1. The shaded squares in Figure MRT1 are holes that has to be cut out.

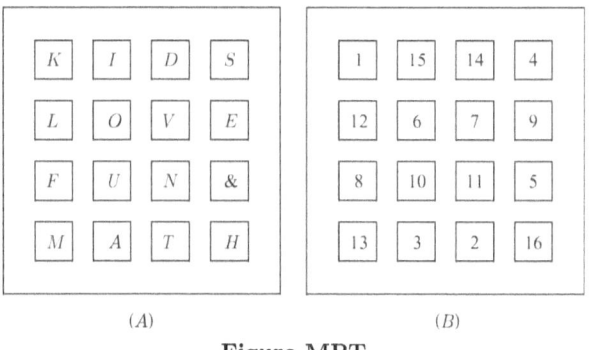

(A)   (B)

Figure MRT

Now ask your spectator to choose a secret letter from card A. Then place one of the cards from Figure MRT1 on top of card A, and ask the spectators whether the secret letter appears through one of the holes. If the answer is "No", then put the top card aside, but you must turn the card over (just as if you are turning a page over in a book; not upside down) If the answer is "Yes", you still have to put the card aside, but do not turn the card over. Repeat this process one by one for all four cards in Figure MRT1. The order of the card is immaterial. Remember to pile all four cards in one pile when you are putting them aside. Then place that pile of four cards (together) on top of card A. Lo and behold! The secret letter is the only one that shows through the hole!

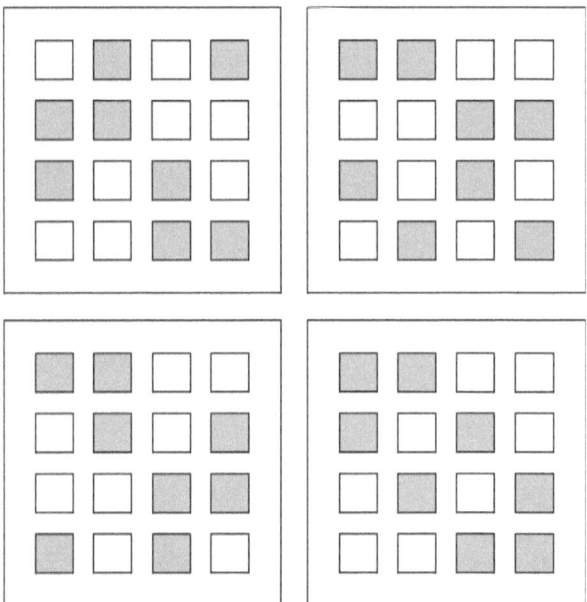

Figure MRT1

From my own experience this trick is always a hit. Spectators are amazed. But why does this trick work? Consider the following. Suppose I ask you the following four questions. Whether the secret letter is in

(a) the right half or left half?

(b) the top half or bottom half?

(c) the even numbered column or the odd numbered column?

(d) the even numbered row or the odd numbered row?

and upon hearing your answers, I guessed your secrete letter correctly. Would you be impressed? Perhaps not; because each question cuts down the number of candidates to half, and after the last question there remains only one candidate. Now suppose we replace the four cards in Figure MRT1 by those in Figure MRT2, then we will be asking essentially the same questions as those listed above. Note that because we are turning the cards over (when the answer is "No"), we can not replace the two cards on the right by the ones obtained by rotating (by 90°) the two cards on the left, unless we agree to turn these two cards upside down when the answer is "No". By now you probably realized that every time we place one of the four cards on top of card A, the number of the remaining candidates is reduced to half just as the four

Mind Reading Tricks

questions above. So, starting from 16 candidates of the secret number, we cut down the number of candidates to 8, 4, 2, and finally to 1. That's all to it.

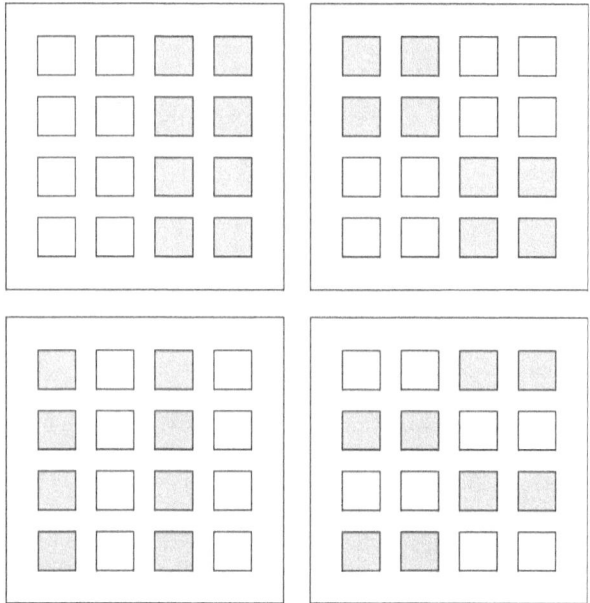

Figure MRT2

We deliberately scattered the holes in the four cards in Figure MRT1 just to mystify the trick, but the idea is exactly the same. Oh yes, I forgot to tell you that instead of card A, you may use any card with 16 symbols or numbers such as card B in Figure MRT.

Exercise.

(a) Can you create your own set of four cards to replace those in Figure MRT1? For example, make holes in two of the four cards symmetric with respect to the center of the card, and those in the other two cards symmetric with respect to the horizontal bisector of the card (but none of the card is the same as those in Figure MRT2).

(b) How many sets of four cards can be constructed?

5.2 Trick 2

Our second trick is based on the same idea, but perhaps more well-known. Again we need five cards. (Figure MRT3)

		A	
1	3	5	7
9	11	13	15
17	19	21	23
25	27	29	31

		B	
2	3	6	7
10	11	14	15
18	19	22	23
26	27	30	31

		C	
4	5	6	7
12	13	14	15
20	21	22	23
28	29	30	31

		D	
8	9	10	11
12	13	14	15
24	25	26	27
28	29	30	31

		E		
16	17	18	19	
20	21	22	23	
24	25	26	27	
28	29	30	31	

Figure MRT3

The audience is requested to think of a secret number from the range 1 through 31 (say, the birthday of the month). Then the performer presents the audience with five cards as in Figure MRT3. As soon as the performer is told by the audience on which of the five cards the secret number appears, the performer declares that he has succeeded in reading the audience's mind and tells the secret number.

How the performer does the trick? At the beginning, there are 31 candidates of the secret number, but there are only 16 numbers on each card. That means that if the secret number appears on the first card, then the number of candidates is reduced to 16, and if it does not appear, then there are only 15 candidates still remaining. In either case, because we still have 4 cards, from our experience in the first trick above, if each card cuts the number of candidates down to half, then it is certainly reasonable to expect that we should be able to pinpoint the secret number.

The question is how? As we have no clue, the best thing we can do is to carry out some experiment to see whether we can find some pattern. Start with small numbers 1 and 2. Clearly, there is only one card for each of them, A and B, respectively. For 3, two cards A and B. For 4, there is only one card C. But for 5, 6, and 7, they are A and C; B and C; A, B, and C, respectively. Recalling that each of 1, 2, 4, appears only on one of cards A, B, C, respectively, we smell something suspicious going on. Namely, if we interpret A, B, C, to be 1, 2, 4, respectively, then it seems that everything fits nicely by doing addition. Let us try a few more cases. The next number is 8, and it appears only in card D. But 9, 10, 11, ..., 15 that follow 8 appear to be just a combination of D and whatever we have before. For example, take 13. Because 13 is 5 more than 8, and 5 appears in cards A and C, if we interpret D to be 8, then everything fits the pattern as before. In fact, we have

$$
\begin{aligned}
9 &= 1 + 8 & (A, D) \\
10 &= 2 + 8 & (B, D) \\
11 &= 1 + 2 + 8 & (A, B, D) \\
12 &= 4 + 8 & (C, D) \\
13 &= 1 + 4 + 8 & (A, C, D) \\
14 &= 2 + 4 + 8 & (B, C, D) \\
15 &= 1 + 2 + 4 + 8 & (A, B, C, D)
\end{aligned}
$$

Mind Reading Tricks

Now we carry on our experiment a few more. And we are confident that we have cracked the secret code. What the performer is doing is just to add the first number on each card that has the secret number written on. Of course, for those who know the binary system, this trick is obvious.

Exercise.

(a) In the first trick, by asking four questions, we are able to pinpoint the secret number out of 16 candidates. But in the second trick, we use five cards to pinpoint the secret one out of only 31 (instead of 32). It appears to be one short. Where is the missing one?

(b) Can you extend the trick? That is, can you expand the range of the secret number (by using more cards)? How about for the first trick also?

(c) In a certain remote village, people do multiplication in the following peculiar way. Take 45×76 for example. Choose one of the factor (it doesn't matter which one, but usually the bigger one, in this case 76), say the one on the right, and multiply repeatedly by 2, while the other factor (the one on the left) is divided by 2 repeatedly, until the quotient becomes 1. Whenever there is a remainder, then simply take the integer part of the quotient. On the other hand, put a notation such as * (just to keep record) if there is no remainder. Finally, add those numbers on the right that does not correspond to * (in our example, $76 + 304 + 608 + 2432 = 3420$), to obtain the answer.

45		76
22	*	152
11		304
5		608
2	*	1216
1		2432
		3420

Can you justify this peculiar multiplication?

Chapter 6

Magic Squares

6.1 New Year Puzzle 2010

Below is a recording of the actual scene when I gave a talk to a group of middle school students in southern California in November 2009.

Without uttering a word, I wrote down on the chalkboard a magic square of order 3 (Figure MGS/Left), and asked, "What is it?" Several students immediately recognized it to be a magic square; i.e., the sum of entries in each row, column, and diagonal is the same. Then I wrote a magic square of order 4 (Figure MGS/Right), and told them the constant sums 15 and 34 are called the *magic sums* for the respective magic squares.

4	9	2
3	5	7
8	1	6

16	2	3	13
5	11	10	8
9	7	6	12
4	14	15	1

Figure MGS

At this point, I asked, "Suppose we have a magic square of order 5 (using numbers 1 through 25), then what would be its magic sum?" After several guesses, I let them bet on their guesses, so they would be committed, hoping that would make them pay more attention to the derivation of the answer. I told them, "I do not know the answer, so let's try to find out." Unfortunately, it turned out that no student guessed the answer correctly, but in the process they learned how to sum an arithmetic progression. Students were amazed that the magic sum can be obtained without actually constructing a magic square.

Then I wrote down a 3×3 grid using the 9 divisors of 36 (Figure MGS2), and asked, "Is it a magic square?" One student noticed it to be a multiplicative magic square; i.e., the product of entries in each row, column, and diagonal is the same.

3	36	2
4	6	9
18	1	12

Figure MGS2

Now, time was ripe for some pep talk. I said, "Let us send a new year puzzle as our season's greetings. It is more fun than the usual cards sold in stores. Everyone can create your own new year puzzle. Here is my example." (See my book, *New Mexico Mathematics Contest Problem Book*, University of New Mexico Press, 2005; Appendix E: New Year Puzzle 1985 - 2016.) Then I asked students, "How many divisors does 2010 have?" As expected, students listed its divisors and counted them. So, I asked, "How do you know there isn't any more?" After a long pause, I had to remind them that all divisors are of the form $2^a \cdot 3^b \cdot 5^c \cdot 67^d$, where a, b, c, d, are either 0 or 1. From this hint, it wasn't hard to figure out how to count the number of divisors from the factorization. "Now, knowing there are 16 divisors, can we arrange them to form a multiplicative magic square? If such a multiplicative magic square does exist, then what should the magic product be?" Unfortunately, no student could make a connection with what they just learned for additive magic square. So I had to remind them that all they needed to do was to change addition to multiplication.

Finally, it was time to construct a new year puzzle. I said, "Of the four prime factors, two of them can be placed in any two cells you like." One student chose to put 5 in the upper left cell. Another student chose the lower right cell for 3. (Figure MGS3/Left) "Now with two factors in the opposite corners, one of the remaining two can be anywhere except on either one of the diagonals." This time a student placed 2 in the cell right next to 5. So I put 67 right above 3 to complete the construction of our New Year Puzzle 2010.

By this time, we had only 5 minutes left. But we managed to fill in the first row and the last column. When the time was up, many students were so excited that they said they would try to complete it themselves.

As can be seen from the description above, I prefer to guide students via "motivational questions". It was obvious that students had fun, and hopefully they saw a way to approach problem-solving.

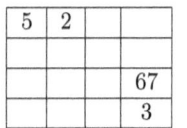

Figure MGS3

Exercises.

(a) What if we put 67 in the fourth row, second column (instead of the third row, fourth column)? (Figure MGS3/Right)

Magic Squares

(b) Where else can we put 67 (assuming we do not change the positions of 2, 3, and 5)?

(c) Suppose there exists a magic square of order n (using numbers 1 through n^2), what should the magic sum be?

(d) Fill in the blanks in the figure below (Figure MGS1) with letters A, B, C, D, such that each letter occurs once and only once in any of the rows, columns or diagonals. How many solutions can you find?

A			
	B		
		C	
			D

Figure MGS1

6.2 Magic Squares

We have two grids (Figure MGS4) with no repetition of letters in each row, column, and diagonal:

A	C	D	B
D	B	A	C
B	D	C	A
C	A	B	D

a	d	b	c
c	b	d	a
d	a	c	b
b	c	a	d

Figure MGS4

Adding these two grids together, we obtain

A + a	C + d	D + b	B + c
D + c	B + b	A + d	C + a
B + d	D + a	C + c	A + b
C + b	A + c	B + a	D + d

Figure MGS5

Because of the way we constructed these grids, in each row, column, and diagonal, each letter appears exactly once (we distinguish upper and lower case letters now), so the sum of each row, column, and diagonal (Figure MGS5) is always

$$(A + B + C + D) + (a + b + c + d).$$

This means that no matter what numbers we substitute into these letters, the sum of each row, column, and diagonal is the same. So, suppose we set

$$\{A, B, C, D\} = \{0, 4, 8, 12\} \quad \text{and} \quad \{a, b, c, d\} = \{1, 2, 3, 4\}.$$

Then we always obtain a magic square! For example, set

$A = 4$, $\quad B = 12$, $\quad C = 0$, $\quad D = 8$, $\quad a = 3$, $\quad b = 4$, $\quad c = 1$, $\quad d = 2$.

Then we obtain the following magic square:

7	2	12	13
9	16	6	3
14	11	1	8
4	5	15	10

Figure MGS6

Exercises.

(a) How many magic squares (using numbers 1 through 16) can you get in this way?

(b) Instead of using
$$\{0, 4, 8, 12\} \quad \text{and} \quad \{1, 2, 3, 4\},$$
for
$$\{A, B, C, D\} \quad \text{and} \quad \{a, b, c, d\},$$
we could use
$$\{0, 2, 4, 6\} \quad \text{and} \quad \{1, 2, 9, 10\};$$
or,
$$\{0, 2, 8, 10\} \quad \text{and} \quad \{1, 2, 5, 6\}.$$
Can you find other such pair of sets? Naturally, sets
$$\{A, B, C, D\} \quad \text{and} \quad \{a, b, c, d\}$$
can be interchanged.

(c) Can this method of creating magic squares of order 4 be applied to magic squares of other sizes?

As for the construction of magic squares of odd order, there are many easy methods. However, the following method does not seem to be well-known. For illustration, we construct a magic square of order 5. We place (5 each of) A, B, C, D, E, and a, b, c, d, e, diagonally as follows (Figure MGS7):

A	E	D	C	B
B	A	E	D	C
C	B	A	E	D
D	C	B	A	E
E	D	C	B	A

b	c	d	e	a
c	d	e	a	b
d	e	a	b	c
e	a	b	c	d
a	b	c	d	e

Figure MGS7

Adding these two grids, we obtain

Magic Squares

A + b	E + c	D + d	C + e	B + a
B + c	A + d	E + e	D + a	C + b
C + d	B + e	A + a	E + b	D + c
D + e	C + a	B + b	A + c	E + d
E + a	D + b	C + c	B + d	A + e

Figure MGS8

Note that the sum of each row or column in the last grid (Figure MGS8) is always

$$(A + B + C + D + E) + (a + b + c + d + e).$$

However, on the two diagonals, we obtain

$$5A + (a + b + c + d + e) \quad \text{and} \quad (A + B + C + D + E) + 5a.$$

Thus, for the sums of the two diagonals to be equal to that of the rows and columns, all we need (i.e., a necessary and sufficient condition) is

$$5A = A + B + C + D + E \quad \text{and} \quad 5a = a + b + c + d + e;$$

i.e., A and a should be the averages of the respective entries. Suppose we choose

$$\{A, B, C, D, E\} = \{1, 2, 3, 4, 5\} \quad \text{and} \quad \{a, b, c, d, e\} = \{0, 5, 10, 15, 20\}.$$

Then, we must have

$$A = 3 \quad \text{and} \quad a = 10.$$

Set the rest as

$$B = 1, \ C = 2, \ D = 5, \ E = 4 \quad \text{and} \quad b = 5, \ c = 20, \ d = 15, \ e = 0.$$

Then we obtain

8	24	20	2	11
21	18	4	15	7
17	1	13	9	25
5	12	6	23	19
14	10	22	16	3

Figure MGS9

Exercises.

(a) How many magic squares (of order 5) can you obtain in this way with a fixed pair of sets $\{A, B, C, D, E\}$ and $\{a, b, c, d, e\}$?

(b) Can you find other pair of sets that works just as

$$\{1, 2, 3, 4, 5\} \quad \text{and} \quad \{0, 5, 10, 15, 20\}?$$

(c) Will this method work for magic squares of even order? Justify your answer.

(d) Note carefully, all the magic squares of odd order created by our method place the average of all the entries at the center cell. Can you find a magic square of odd order that does not have this property?

Chapter 7

Fun with Areas

7.1 Two Theorems of Newton

Theorem. [Newton] Suppose a quadrangle $ABCD$ is circumscribing a circle $O(r)$. (Figure CLP) Then the midpoints M, N, of diagonals AC, BD, and center O (of the circle) are collinear.

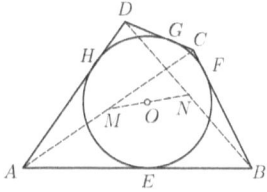

Figure CLP

A rather conspicuous property common to these three points is that regardless of whether $P = M$, N, or O, we have

$$[PAB] + [PCD] = \frac{1}{2}[ABCD] = [PBC] + [PDA].$$

Because M is the midpoint of AC, we have

$$[MAB] + [MCD] = \frac{1}{2}([CAB] + [ACD]) = \frac{1}{2}[ABCD].$$

Hence

$$[MBC] + [MDA] = \frac{1}{2}[ABCD];$$

and similarly,

$$[NAB] + [NCD] = \frac{1}{2}[ABCD] = [NBC] + [NDA].$$

Let E, F, G, H be the points where circle $O(r)$ touches sides AB, BC, CD, DA, respectively. Then clearly, we have

$$\overline{AH} = \overline{AE}, \quad \overline{BE} = \overline{BF}, \quad \overline{CF} = \overline{CG}, \quad \overline{DG} = \overline{DH}.$$

Hence

$$\begin{aligned}\overline{AB} + \overline{CD} &= (\overline{AE} + \overline{BE}) + (\overline{CG} + \overline{DG}) \\ &= \overline{AH} + (\overline{BF} + \overline{CF}) + \overline{DH} = \overline{BC} + \overline{DA}.\end{aligned}$$

Therefore,

$$\begin{aligned}[OAB] + [OCD] &= \frac{r}{2}(\overline{AB} + \overline{CD}) \\ &= \frac{r}{2}(\overline{BC} + \overline{DA}) = [OBC] + [ODA].\end{aligned}$$

This observation makes us suspicious whether the locus of points P satisfying

$$[PAB] + [PCD] = \text{ a constant}$$

is a straight line. Luckily, this conjecture turns out to be not only correct, but also easy to establish.

We adopt a convention that the area $[XYZ]$ of $\triangle XYZ$ is positive if X-Y-Z is labelled counterclockwise, but negative if it is clockwise. (Figure CLP1) (Naturally, $[XYZ] = 0$ if $\triangle XYZ$ degenerates; i.e., if X, Y, Z are collinear.)

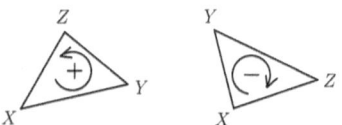

Figure CLP1

With this convention, we have the following.

Lemma. Given two non-parallel segments AB and CD in a plane, the locus of points P satisfying

$$[PAB] + [PCD] = \text{ a constant}$$

is a straight line. (Figure CLP2/Left)

Fun with Areas

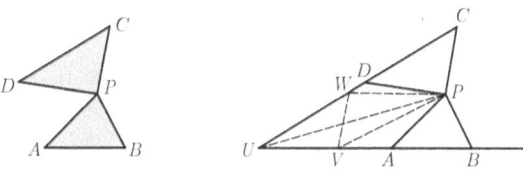

Figure CLP2

Proof. Because $[PAB] = [PA'B']$ for any segment $A'B'$ on line AB with the same length as AB, we might as well choose one that is easiest for our purpose. (Figure CLP2/Right) Let U be the intersection of lines AB and CD. Choose V, W on lines AB and CD, respectively, so that $\overrightarrow{UV} = \overrightarrow{AB}$, and $\overrightarrow{WU} = \overrightarrow{CD}$. Then, clearly,

$$[PAB] + [PCD] = [PUV] + [PWU] = [UVW] + [PWV].$$

Now, $[UVW] = $ a constant, and so the condition

$$[PAB] + [PCD] = \text{a constant}$$

becomes $[PWV] = $ a constant. But the locus of the points P that satisfy this condition is clearly a line parallel to VW. □

Remark. Even if $AB \parallel CD$, as long as $\overline{AB} \neq \overline{CD}$, then the conclusion is still true. What if $\overline{AB} = \overline{CD}$?

Once the lemma is established, then our proof of the theorem at the beginning is complete. (The case $AB \parallel CD$ or $BC \parallel DA$ is trivial.) But is there other application of the lemma?

Theorem. [Newton] Suppose four straight lines intersect pairwise at six points, A, B, C, D, E, F. (Figure CLP3) Then the midpoints L, M, N, of the "diagonals" AC, BD, EF, are collinear.

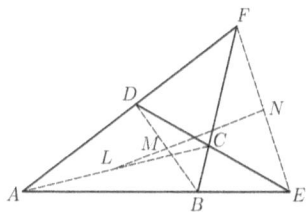

Figure CLP3

Proof. By lemma, all we need is to show that all three points L, M, N satisfy the condition that

$$[PAB] + [PCD] = \text{a constant}.$$

But this is immediate.

$$[LAB] + [LCD] = \frac{1}{2}([CAB] + [ACD]) = \frac{1}{2}[ABCD],$$
$$[MAB] + [MCD] = \frac{1}{2}([DAB] + [BCD]) = \frac{1}{2}[ABCD],$$
$$[NAB] + [NCD] = \frac{1}{2}([FAB] + [FCD])$$
$$= \frac{1}{2}([FAB] - [FDC]) = \frac{1}{2}[ABCD].$$

Hence our proof is complete. □

Remark. Given non-parallel segments AB and CD in a plane such that $\overline{AB} = \overline{CD}$, then two lines

$$[PAB] + [PCD] = \text{a constant} \quad \text{and} \quad [QAB] - [QCD] = \text{a constant}$$

are perpendicular. (Why?) I would appreciate it if the reader would let me know an application of this extension of the lemma.

Now we want to generalize the Newton theorem using determinants with complex numbers. We start with preparatory comments. It is well-known that given three points

$$X(x = x_1 + ix_2), \ Y(y = y_1 + iy_2), \ Z(z = z_1 + iz_2)$$

(where x_1, x_2, y_1, y_2, z_1, z_2 are real numbers), the signed area of $\triangle XYZ$ is

$$[XYZ] = \frac{1}{2} \begin{vmatrix} x_1 & x_2 & 1 \\ y_1 & y_2 & 1 \\ z_1 & z_2 & 1 \end{vmatrix}.$$

Using the relations

$$x_1 = \frac{x + \bar{x}}{2}, \ x_2 = \frac{x - \bar{x}}{2i}, \ \text{etc.},$$

we obtain easily the area formula in terms of complex numbers as

$$[XYZ] = \frac{i}{4} \begin{vmatrix} x & \bar{x} & 1 \\ y & \bar{y} & 1 \\ z & \bar{z} & 1 \end{vmatrix}.$$

Next, we need the following determinant identity. For any six complex numbers p, q, r, u, v, w, we claim

$$\begin{vmatrix} p+u & \bar{p}+\bar{u} & 1 \\ q+v & \bar{q}+\bar{v} & 1 \\ r+w & \bar{r}+\bar{w} & 1 \end{vmatrix} = \begin{vmatrix} p & \bar{p} & 1 \\ q & \bar{q} & 1 \\ r & \bar{r} & 1 \end{vmatrix} + \begin{vmatrix} p & \bar{p} & 1 \\ v & \bar{v} & 1 \\ w & \bar{w} & 1 \end{vmatrix} + \begin{vmatrix} q & \bar{q} & 1 \\ w & \bar{w} & 1 \\ u & \bar{u} & 1 \end{vmatrix} + \begin{vmatrix} r & \bar{r} & 1 \\ u & \bar{u} & 1 \\ v & \bar{v} & 1 \end{vmatrix}.$$

Fun with Areas

Proof is by a straightforward computation. By the multilinearity of determinants,

$$\begin{vmatrix} p+u & \bar{p}+\bar{u} & 1 \\ q+v & \bar{q}+\bar{v} & 1 \\ r+w & \bar{r}+\bar{w} & 1 \end{vmatrix} = \begin{vmatrix} p & \bar{p} & 1 \\ q & \bar{q} & 1 \\ r & \bar{r} & 1 \end{vmatrix} + \begin{vmatrix} p & \bar{u} & 1 \\ q & \bar{v} & 1 \\ r & \bar{w} & 1 \end{vmatrix} + \begin{vmatrix} u & \bar{p} & 1 \\ v & \bar{q} & 1 \\ w & \bar{r} & 1 \end{vmatrix} + \begin{vmatrix} u & \bar{u} & 1 \\ v & \bar{v} & 1 \\ w & \bar{w} & 1 \end{vmatrix}.$$

As the first determinant on the right-hand side appears in the final result, we shall keep it as it is. And we expand the remaining three determinants; then reassemble them to form three new determinants. To do so, we expand the first determinant (to be expanded) with respect to the first column; the second determinant (to be expanded) with respect to the second column; the third determinant (to be expanded) with respect to the third column. Then we collect the first terms of the expansions to form a new determinant. Similarly, we collect the second terms of the expansions to form another determinant; and we collect the third terms of the expansions to form yet another determinant. Namely,

$$\begin{vmatrix} p & \bar{u} & 1 \\ q & \bar{v} & 1 \\ r & \bar{w} & 1 \end{vmatrix} + \begin{vmatrix} u & \bar{p} & 1 \\ v & \bar{q} & 1 \\ w & \bar{r} & 1 \end{vmatrix} + \begin{vmatrix} u & \bar{u} & 1 \\ v & \bar{v} & 1 \\ w & \bar{w} & 1 \end{vmatrix}$$

$$= \left\{ p\begin{vmatrix} \bar{v} & 1 \\ \bar{w} & 1 \end{vmatrix} - \bar{p}\begin{vmatrix} v & 1 \\ w & 1 \end{vmatrix} + \begin{vmatrix} v & \bar{v} \\ w & \bar{w} \end{vmatrix} \right\}$$

$$+ \left\{ q\begin{vmatrix} \bar{w} & 1 \\ \bar{u} & 1 \end{vmatrix} - \bar{q}\begin{vmatrix} w & 1 \\ u & 1 \end{vmatrix} + \begin{vmatrix} w & \bar{w} \\ u & \bar{u} \end{vmatrix} \right\}$$

$$+ \left\{ r\begin{vmatrix} \bar{u} & 1 \\ \bar{v} & 1 \end{vmatrix} - \bar{r}\begin{vmatrix} u & 1 \\ v & 1 \end{vmatrix} + \begin{vmatrix} u & \bar{u} \\ v & \bar{v} \end{vmatrix} \right\}$$

$$= \begin{vmatrix} p & \bar{p} & 1 \\ v & \bar{v} & 1 \\ w & \bar{w} & 1 \end{vmatrix} + \begin{vmatrix} q & \bar{q} & 1 \\ w & \bar{w} & 1 \\ u & \bar{u} & 1 \end{vmatrix} + \begin{vmatrix} r & \bar{r} & 1 \\ u & \bar{u} & 1 \\ v & \bar{v} & 1 \end{vmatrix}.$$

Thus we obtained the desired determinant identity. And we have the following generalization of the Newton theorem immediately.

Lemma. For any six points P, Q, R, U, V, W, in a plane, let L, M, N, be the midpoints of segments PU, QV, RW, respectively, then (with signed areas),

$$\begin{aligned} 4[LMN] &= [PQR] + [PVW] + [QWU] + [RUV] \\ &= [UVW] + [UQR] + [VRP] + [WPQ]. \end{aligned}$$

Proof. Set $P(p)$, $Q(q)$, $R(r)$, $U(u)$, $V(v)$, $W(w)$. Because $L\left(\frac{p+u}{2}\right)$, $M\left(\frac{q+v}{2}\right)$, $N\left(\frac{r+w}{2}\right)$, we have

$$[LMN] = \frac{i}{4} \begin{vmatrix} \frac{p+u}{2} & \frac{\bar{p}+\bar{u}}{2} & 1 \\ \frac{q+v}{2} & \frac{\bar{q}+\bar{v}}{2} & 1 \\ \frac{r+w}{2} & \frac{\bar{r}+\bar{w}}{2} & 1 \end{vmatrix} = \frac{1}{4} \cdot \frac{i}{4} \begin{vmatrix} p+u & \bar{p}+\bar{u} & 1 \\ q+v & \bar{q}+\bar{v} & 1 \\ r+w & \bar{r}+\bar{w} & 1 \end{vmatrix}$$

and the determinant identity above finishes the proof. $\square$

To show that the Newton theorem is a particular case of this lemma, we note that in Figure CLP3 (page 55), points L, M, N, are the midpoints of AC, BD, EF, respectively; hence let points A, B, E, C, D, F, correspond to P, Q, R, U, V, W, respectively, in the lemma. We obtain

$$4[LMN] = [ABE] + [ADF] + [BFC] + [ECD].$$

But each of the four triangles on the right-hand side degenerates (to a line segment), and so the right-hand side is zero, implying that points L, M, N, are collinear.

Had we used the other identity, we still obtain the same result.

$$\begin{aligned} 4[LMN] &= [CDF] + [CBE] + [DEA] + [FAB] \\ &= [CDF] + \{[CBE] + [DEA]\} + [FAB] \\ &= \{[CDF] + [DCBA]\} + [FAB] = [FBA] + [FAB] = 0. \end{aligned}$$

Exercises.

1. Let ABC be a right triangle with $\angle C = 90°$.

   (a) Suppose P, Q, R, are points such that
   $$\triangle PBC \sim \triangle QCA \sim \triangle RBA \quad (\text{not } \triangle RAB);$$
   and L, M, N, are the midpoints of AP, BQ, CR, respectively. Show that
   $$4[LMN] = [PQR].$$

   (b) Let H be the foot of the perpendicular from vertex C to the hypotenuse AB; P, Q, the mirror images of point H with respect to legs BC, CA, respectively; and L, M, are the midpoints of AP, BQ, respectively. Show that points L, M, C, are collinear.

2. Show that, for any nine points P, Q, R, U, V, W, X, Y, Z, we have
$$\begin{aligned} 9[LMN] &= [PQR] + [PVW] + [QWU] + [RUV] \\ &\quad + [PYZ] + [QZX] + [RXY] \\ &= [UVW] + [UYZ] + [VZX] + [WXY] \\ &\quad + [UQR] + [VRP] + [WPQ] \\ &= [XYZ] + [XQR] + [YRP] + [ZPQ] \\ &\quad + [XVW] + [YWU] + [ZUV], \end{aligned}$$

Fun with Areas

where L, M, N, are the centroids of $\triangle PUX$, $\triangle QVY$, $\triangle RWZ$, respectively.

3. Can you extend the lemma to the three dimensional case?

7.2 A Charming Construction Problem

The following problem is communicated to me by Mr. Dean Ballard of Lakeside School, Seattle. Although it is elementary, it has a charming solution that I would like to offer readers a chance to try it yourselves.[1]

> Given an equilateral triangle ABC, construct points P on side BC; Q on side CA; R on side AB; such that $\overline{PB} = \overline{QC} = \overline{RA}$, and the area of triangle PQR is one half of that of triangle ABC. (Figure CCP)

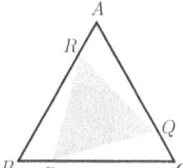

Figure CCP

As a byproduct of solving the problem above, I came up with the following.

> **Theorem.** Given triangle ABC, let U, V, W be the midpoints of sides BC, CA, AB, respectively. (CCP1) Suppose P, P' are points on side BC (or its extension); Q, Q' are points on side CA (or its extension); R, R' are points on side AB (or its extension); such that
> $$\overline{UP} = \overline{UP'}, \quad \overline{VQ} = \overline{VQ'}, \quad \overline{WR} = \overline{WR'}.$$
> Then
> $$[PQR] = [P'Q'R'].$$

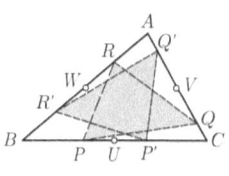 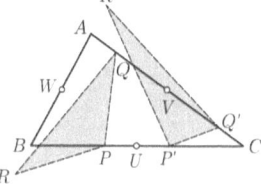

Figure CCP1

---

[1] For a similar problem, See my book, *Honsberger Revisited* (National Taiwan University Press, Taipei, 2012), Exercise (e), page 223.

Proof. Instead of $A(\alpha)$, $B(\beta)$, $C(\gamma)$, we may use $U(u)$, $V(v)$, $W(w)$, where of course $u = \frac{1}{2}(\beta + \gamma)$, $v = \frac{1}{2}(\gamma + \alpha)$, $w = \frac{1}{2}(\alpha + \beta)$. Because $BC \parallel VW$, point P on line (not just on segment!) BC can be represented by a complex number $u + p(v - w)$, where parameter p is a suitable real number. Then point P' can be represented by $u - p(v - w)$. Similarly, points Q and R can be represented by $v + q(w - u)$ and $w + r(u - v)$, respectively, where parameters q and r are some suitable real numbers; and for points Q' and R', simply change the signs of parameters q and r. Therefore,

$$[PQR] = \frac{i}{4} \begin{vmatrix} u + p(v - w) & \bar{u} + p(\bar{v} - \bar{w}) & 1 \\ v + q(w - u) & \bar{v} + q(\bar{w} - \bar{u}) & 1 \\ w + r(u - v) & \bar{w} + r(\bar{u} - \bar{v}) & 1 \end{vmatrix}.$$

We want to show that $[PQR] = [P'Q'R']$; i.e., the value of this determinant remains unchanged when all three signs of parameters p, q, r are changed simultaneously.

To evaluate the value of this determinant, we use the multilinearity of determinants again.

$$\begin{vmatrix} u + p(v - w) & \bar{u} + p(\bar{v} - \bar{w}) & 1 \\ v + q(w - u) & \bar{v} + q(\bar{w} - \bar{u}) & 1 \\ w + r(u - v) & \bar{w} + r(\bar{u} - \bar{v}) & 1 \end{vmatrix} = \begin{vmatrix} u & \bar{u} & 1 \\ v & \bar{v} & 1 \\ w & \bar{w} & 1 \end{vmatrix} + \begin{vmatrix} p(v - w) & p(\bar{v} - \bar{w}) & 1 \\ q(w - u) & q(\bar{w} - \bar{u}) & 1 \\ r(u - v) & r(\bar{u} - \bar{v}) & 1 \end{vmatrix}$$
$$+ \begin{vmatrix} p(v - w) & \bar{u} & 1 \\ q(w - u) & \bar{v} & 1 \\ r(u - v) & \bar{w} & 1 \end{vmatrix} + \begin{vmatrix} u & p(\bar{v} - \bar{w}) & 1 \\ v & q(\bar{w} - \bar{u}) & 1 \\ w & r(\bar{u} - \bar{v}) & 1 \end{vmatrix}.$$

Obviously, the sum of the last two determinants is purely imaginary. Hence the coefficients of the terms such as $u\bar{u}$ must be zero (in fact, $q + r - q - r = 0$); and the coefficients of terms such as $v\bar{w}$ and $\bar{v}w$ must differ by signs only. But the coefficients of $v\bar{w}$ is $-p + p = 0$, which is symmetric with respect to p, q, r. And so we conclude that the sum of the last two determinants is zero. We obtain

$$-4i[PQR] = \begin{vmatrix} u & \bar{u} & 1 \\ v & \bar{v} & 1 \\ w & \bar{w} & 1 \end{vmatrix} + \begin{vmatrix} p(v - w) & p(\bar{v} - \bar{w}) & 1 \\ q(w - u) & q(\bar{w} - \bar{u}) & 1 \\ r(u - v) & r(\bar{u} - \bar{v}) & 1 \end{vmatrix},$$

which is invariant under the change of signs of parameters p, q, r, simultaneously. Therefore, we obtain the desired result

$$[PQR] = [P'Q'R'].$$

However, let us evaluate the second determinant on the right-hand side. Again, it is obvious that its value is purely imaginary. So the coefficients of terms such as $u\bar{u}$ must be zero (in fact, $-qr + qr = 0$); and the coefficients of terms such as $v\bar{w}$ and $\bar{v}w$ must differ by signs only. It turns out that the coefficient of $v\bar{w}$ is $qr + rp + pq$, which is symmetric with respect to p, q, r. Hence we conclude that

$$\begin{vmatrix} p(v - w) & p(\bar{v} - \bar{w}) & 1 \\ q(w - u) & q(\bar{w} - \bar{u}) & 1 \\ r(u - v) & r(\bar{u} - \bar{v}) & 1 \end{vmatrix}$$

Fun with Areas

$$= (qr + rp + pq) \cdot [(v\bar{w} - \bar{v}w) + (w\bar{u} - \bar{w}u) + (u\bar{v} - \bar{u}v)]$$

$$= (qr + rp + pq) \cdot \begin{vmatrix} u & \bar{u} & 1 \\ v & \bar{v} & 1 \\ w & \bar{w} & 1 \end{vmatrix}$$

$$= -4i(qr + rp + pq)[UVW].$$

$$[PQR] = (qr + rp + pq + 1)[UVW] = \frac{1}{4}(qr + rp + pq + 1)[ABC].$$

□

Alternate Proof. The following proof, which allows us to generalize the theorem, is communicated to me by Mr. Tom Rona of Lakeside School, Seattle, while he was vacationing is Paris, in July 2012.

Again, we use complex numbers. Set

$$A(\alpha), \ B(\beta), \ C(\gamma), \ P(p), \ P'(p'), \ Q(q), \ Q'(q'), \ R(r), \ R'(r').$$

Because side BC and segment PP' share the common midpoint, we have

$$\beta + \gamma = p + p', \ \gamma + \alpha = q + q', \ \alpha + \beta = r + r'.$$

It follow that $p' = \beta + \gamma - p$, $q' = \gamma + \alpha - q$, $r' = \alpha + \beta - r$.
Therefore,

$$-4i[P'Q'R'] = \begin{vmatrix} p' & \bar{p}' & 1 \\ q' & \bar{q}' & 1 \\ r' & \bar{r}' & 1 \end{vmatrix} = \begin{vmatrix} \beta+\gamma-p & \bar{\beta}+\bar{\gamma}-\bar{p} & 1 \\ \gamma+\alpha-q & \bar{\gamma}+\bar{\alpha}-\bar{q} & 1 \\ \alpha+\beta-r & \bar{\alpha}+\bar{\beta}-\bar{r} & 1 \end{vmatrix}$$

$$= \begin{vmatrix} p & \bar{p} & 1 \\ q & \bar{q} & 1 \\ r & \bar{r} & 1 \end{vmatrix} - \begin{vmatrix} p & \bar{\beta}+\bar{\gamma} & 1 \\ q & \bar{\gamma}+\bar{\alpha} & 1 \\ r & \bar{\alpha}+\bar{\beta} & 1 \end{vmatrix} - \begin{vmatrix} \beta+\gamma & \bar{p} & 1 \\ \gamma+\alpha & \bar{q} & 1 \\ \alpha+\beta & \bar{r} & 1 \end{vmatrix}$$

$$+ \begin{vmatrix} \beta+\gamma & \bar{\beta}+\bar{\gamma} & 1 \\ \gamma+\alpha & \bar{\gamma}+\bar{\alpha} & 1 \\ \alpha+\beta & \bar{\alpha}+\bar{\beta} & 1 \end{vmatrix}$$

$$= -4i[PQR]$$

$$- p \begin{vmatrix} \bar{\gamma}+\bar{\alpha} & 1 \\ \bar{\alpha}+\bar{\beta} & 1 \end{vmatrix} + \bar{p} \begin{vmatrix} \gamma+\alpha & 1 \\ \alpha+\beta & 1 \end{vmatrix} + \begin{vmatrix} \gamma+\alpha & \bar{\gamma}+\bar{\alpha} \\ \alpha+\beta & \bar{\alpha}+\bar{\beta} \end{vmatrix}$$

$$- q \begin{vmatrix} \bar{\alpha}+\bar{\beta} & 1 \\ \bar{\beta}+\bar{\gamma} & 1 \end{vmatrix} + \bar{q} \begin{vmatrix} \alpha+\beta & 1 \\ \beta+\gamma & 1 \end{vmatrix} + \begin{vmatrix} \alpha+\beta & \bar{\alpha}+\bar{\beta} \\ \beta+\gamma & \bar{\beta}+\bar{\gamma} \end{vmatrix}$$

$$- r \begin{vmatrix} \bar{\beta}+\bar{\gamma} & 1 \\ \bar{\gamma}+\bar{\alpha} & 1 \end{vmatrix} + \bar{r} \begin{vmatrix} \beta+\gamma & 1 \\ \gamma+\alpha & 1 \end{vmatrix} + \begin{vmatrix} \beta+\gamma & \bar{\beta}+\bar{\gamma} \\ \gamma+\alpha & \bar{\gamma}+\bar{\alpha} \end{vmatrix}$$

$$= -4i[PQR] + \{p(\bar{\beta}-\bar{\gamma}) - \bar{p}(\beta-\gamma)$$
$$+ [-(\beta\bar{\gamma}-\bar{\beta}\gamma) + (\gamma\bar{\alpha}-\bar{\gamma}\alpha) + (\alpha\bar{\beta}-\bar{\alpha}\beta)]\}$$

$$+ \{q(\bar{\gamma} - \bar{\alpha}) - \bar{q}(\gamma - \alpha)$$
$$+ [(\beta\bar{\gamma} - \bar{\gamma}\alpha) - (\gamma\bar{\alpha} - \bar{\gamma}\alpha) + (\alpha\bar{\beta} - \bar{\alpha}\beta)]\}$$
$$+ \{r(\bar{\alpha} - \bar{\beta}) - \bar{r}(\alpha - \beta)$$
$$+ [(\beta\bar{\gamma} - \bar{\beta}\gamma) + (\gamma\bar{\alpha} - \bar{\gamma}\alpha) - (\alpha\bar{\beta} - \bar{\alpha}\beta)]\}$$
$$= -4i[PQR] + \{p(\bar{\beta} - \bar{\gamma}) - \bar{p}(\beta - \gamma) + (\beta\bar{\gamma} - \bar{\beta}\gamma)\}$$
$$+ \{q(\bar{\gamma} - \bar{\alpha}) - \bar{q}(\gamma - \alpha) + (\gamma\bar{\alpha} - \bar{\gamma}\alpha)\}$$
$$+ \{r(\bar{\alpha} - \bar{\beta}) - \bar{r}(\alpha - \beta) + (\alpha\bar{\beta} - \bar{\alpha}\beta)\}$$
$$= -4i[PQR] + \begin{vmatrix} p & \bar{p} & 1 \\ \beta & \bar{\beta} & 1 \\ \gamma & \bar{\gamma} & 1 \end{vmatrix} + \begin{vmatrix} q & \bar{q} & 1 \\ \gamma & \bar{\gamma} & 1 \\ \alpha & \bar{\alpha} & 1 \end{vmatrix} + \begin{vmatrix} r & \bar{r} & 1 \\ \alpha & \bar{\alpha} & 1 \\ \beta & \bar{\beta} & 1 \end{vmatrix}$$
$$= -4i\{[PQR] + [PBC] + [QCA] + [RAB]\}$$
$$= -4i[PQR].$$

because points P, B, C are collinear; so are points Q, C, A; and points R, A, B. We have shown that
$$[P'Q'R'] = [PQR].$$

Corollary. Given $\triangle ABC$, let $BPCP'$, $CQAQ'$, $ARBR'$ be parallelograms, then (with signed areas),
$$[P'Q'R'] = [PQR] + [PBC] + [QCA] + [RAB]$$

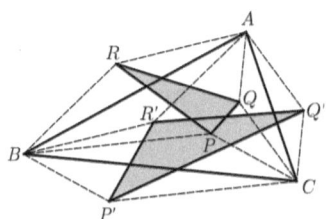

Figure CCP2

7.3 A Generalization of the Simson Theorem

Let D be a variable point and P, Q, R, the feet of perpendiculars from point D to sides BC, CA, AB (or their extensions) of $\triangle ABC$. What is the locus of point D such that the area of the *pedal triangle* PQR associated with point D (with respect to $\triangle ABC$) is constant? (Figure GST)

Fun with Areas

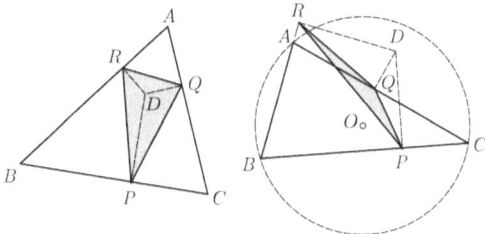

Figure GST

Again we work on the complex plane. Without loss of generality, we may assume that $\triangle ABC$ is inscribed in the unit circle; i.e., let

$$A(\alpha), \quad B(\beta), \quad C(\gamma), \quad D(\delta), \quad \text{where } |\alpha| = |\beta| = |\gamma| = 1.$$

First, we want to express feet $P(\lambda)$, $Q(\mu)$, $R(\nu)$, in terms of α, β, γ, and δ. Thus we want to find the equations of lines BC and DP, as $P(\lambda)$ is the intersection of these two lines.

Because vector $\zeta = \beta + \gamma$ is perpendicular to line BC, for an arbitrary point z on line BC,

$$\frac{z - \beta}{\zeta} \text{ must be purely imaginary}$$

$$\iff \frac{z - \beta}{\zeta} + \frac{\bar{z} - \bar{\beta}}{\bar{\zeta}} = 0$$

$$\iff \frac{z}{\zeta} + \frac{\bar{z}}{\bar{\zeta}} = \frac{\beta}{\zeta} + \frac{\bar{\beta}}{\bar{\zeta}}$$

$$\iff z + k\bar{z} = \beta + k\bar{\beta} \quad (k = \frac{\zeta}{\bar{\zeta}} = \beta\gamma, \text{ because } \bar{\beta} = \frac{1}{\beta}, \bar{\gamma} = \frac{1}{\gamma})$$

$$\iff z + \beta\gamma\bar{z} = \beta + \gamma.$$

The last equality is an equation of line BC. (Note that if $\zeta = \beta + \gamma = 0$, then simply replace ζ by a vector perpendicular to line BC, say $i\beta$, in the computation above, and we still obtain the same equation of line BC as above. Same comment for the next computation.)

Similarly, for any point z on line DP (because vector $i\zeta$ is perpendicular to line DP), we have

$$\frac{z - \delta}{i\zeta} \text{ must be purely imaginary}$$

$$\iff \frac{z - \delta}{i\zeta} - \frac{\bar{z} - \bar{\delta}}{\overline{i\zeta}} = 0$$

$$\iff \frac{z}{\zeta} - \frac{\bar{z}}{\bar{\zeta}} = \frac{\delta}{\zeta} - \frac{\bar{\delta}}{\bar{\zeta}}$$

$$\iff z - k\bar{z} = \delta - k\bar{\delta}$$
$$\iff z - \beta\gamma\bar{z} = \delta - \beta\gamma\bar{\delta}.$$

The last equality is an equation of line DP. So to obtain feet $P(\lambda)$ of the perpendicular from point $D(\delta)$ to line BC, all we have to do is to solve the simultaneous equations. It is immediate that

$$\lambda = \frac{1}{2}(\beta + \gamma + \delta - \beta\gamma\bar{\delta}).$$

Similarly, we obtain

$$\mu = \frac{1}{2}(\gamma + \alpha + \delta - \gamma\alpha\bar{\delta}),$$
$$\nu = \frac{1}{2}(\alpha + \beta + \delta - \alpha\beta\bar{\delta}).$$

Now we are ready to compute the area of $\triangle PQR$.

$$\begin{aligned}
[PQR] &= \frac{i}{4}\begin{vmatrix} \lambda & \bar{\lambda} & 1 \\ \mu & \bar{\mu} & 1 \\ \nu & \bar{\nu} & 1 \end{vmatrix} \\
&= \frac{i}{4^2}\begin{vmatrix} \beta+\gamma+\delta-\beta\gamma\bar{\delta} & \bar{\beta}+\bar{\gamma}+\bar{\delta}-\bar{\beta}\bar{\gamma}\delta & 1 \\ \gamma+\alpha+\delta-\gamma\alpha\bar{\delta} & \bar{\gamma}+\bar{\alpha}+\bar{\delta}-\bar{\gamma}\bar{\alpha}\delta & 1 \\ \alpha+\beta+\delta-\alpha\beta\bar{\delta} & \bar{\alpha}+\bar{\beta}+\bar{\delta}-\bar{\alpha}\bar{\beta}\delta & 1 \end{vmatrix} \\
&= \frac{i}{4^2}\begin{vmatrix} \gamma-\alpha-\beta\bar{\delta}(\gamma-\alpha) & \bar{\gamma}-\bar{\alpha}-\bar{\beta}\delta(\bar{\gamma}-\bar{\alpha}) & 0 \\ \gamma+\alpha+\delta-\gamma\alpha\bar{\delta} & \bar{\gamma}+\bar{\alpha}+\bar{\delta}-\bar{\gamma}\bar{\alpha}\delta & 1 \\ \alpha+\beta+\delta-\alpha\beta\bar{\delta} & \bar{\alpha}+\bar{\beta}+\bar{\delta}-\bar{\alpha}\bar{\beta}\delta & 1 \end{vmatrix} \\
&= \frac{i}{4^2}\begin{vmatrix} \gamma-\alpha-\beta\bar{\delta}(\gamma-\alpha) & \bar{\gamma}-\bar{\alpha}-\bar{\beta}\delta(\bar{\gamma}-\bar{\alpha}) & 0 \\ \gamma+\alpha+\delta-\gamma\alpha\bar{\delta} & \bar{\gamma}+\bar{\alpha}+\bar{\delta}-\bar{\gamma}\bar{\alpha}\delta & 1 \\ \beta-\gamma-\alpha\bar{\delta}(\beta-\gamma) & \bar{\beta}-\bar{\gamma}-\bar{\alpha}\delta(\bar{\beta}-\bar{\gamma}) & 0 \end{vmatrix} \\
&= -\frac{i}{4^2}\begin{vmatrix} (\gamma-\alpha)(1-\beta\bar{\delta}) & (\bar{\gamma}-\bar{\alpha})(1-\bar{\beta}\delta) \\ (\beta-\gamma)(1-\alpha\bar{\delta}) & (\bar{\beta}-\bar{\gamma})(1-\bar{\alpha}\delta) \end{vmatrix} \quad (\delta\bar{\delta}=r^2) \\
&= -\frac{i}{4^2}\begin{vmatrix} (\gamma-\alpha)\cdot\frac{\beta r^2-\delta}{\delta} & \frac{\gamma-\alpha}{\gamma\alpha}\cdot\frac{\beta-\delta}{\beta} \\ (\beta-\gamma)\cdot\frac{\alpha r^2-\delta}{\delta} & \frac{\beta-\gamma}{\beta\gamma}\cdot\frac{\alpha-\delta}{\alpha} \end{vmatrix} \\
&= -\frac{i(\beta-\gamma)(\gamma-\alpha)}{4^2\alpha\beta\gamma\delta}\begin{vmatrix} \beta r^2-\delta & \beta-\delta \\ \alpha r^2-\delta & \alpha-\delta \end{vmatrix} \\
&= -\frac{i(\beta-\gamma)(\gamma-\alpha)}{4^2\alpha\beta\gamma\delta}\begin{vmatrix} \beta r^2-\delta & \beta-\delta \\ (\alpha-\beta)r^2 & \alpha-\beta \end{vmatrix} \\
&= -\frac{i(\beta-\gamma)(\gamma-\alpha)(\alpha-\beta)}{4^2\alpha\beta\gamma\delta}\begin{vmatrix} \beta r^2-\delta & \beta-\delta \\ r^2 & 1 \end{vmatrix} \\
&= -\frac{i(\beta-\gamma)(\gamma-\alpha)(\alpha-\beta)(r^2-1)}{4^2\alpha\beta\gamma}
\end{aligned}$$

Fun with Areas

$$= \frac{1-r^2}{4} \cdot \frac{i}{4} \begin{vmatrix} \alpha & \bar{\alpha} & 1 \\ \beta & \bar{\beta} & 1 \\ \gamma & \bar{\gamma} & 1 \end{vmatrix}$$

$$= \frac{1-r^2}{4}[ABC].$$

It follows that

points P, Q, R are collinear
$\Longleftrightarrow$ the area of pedal trangle PQR is zero
$\Longleftrightarrow$ $r = 1$ (assuming that $\triangle ABC$ is not degernate)
$\Longleftrightarrow$ point D is on the circumcircle of $\triangle ABC$.

Observe that the result of our computation above shows that the *signed* area of the pedal triangle associated with a variable point D (with respect to $\triangle ABC$) depends only on the distance r of point D from the circumcircle of $\triangle ABC$. Hence we obtain the following.

Theorem. (a) The locus of point D that keeps the *signed* area of its pedal triangle PQR (with respect to $\triangle ABC$) constant is a circle concentric with the circumcircle of $\triangle ABC$. In particular, feet P, Q, R of the perpendiculars from point D to the sides (or their extensions) are collinear if and only if D is on the circumcircle of $\triangle ABC$.
(b) Suppose the signed area of $\triangle ABC$ is positive. Then, of all the pedal triangles PRQ (with respect to $\triangle ABC$), the one associated with the circumcenter of $\triangle ABC$ has the maximum signed area, which is $\frac{1}{4}$ of the area of $\triangle ABC$.

Remark. For more generalizations of the Simson theorem, see my book, *Complex Numbers and Geometry* (Mathematical Association of America, 1994).

Chapter 8

The Tower of Hanoi

The Tower of Hanoi consists of eight (or rather, n) wooden discs of different sizes, and three pegs fastened to a stand. Each of the discs has a hole in the middle so that a peg can be passed. Initially, the discs are placed all on one peg with the smallest at the top, and the sizes of the successive discs increase as we descend; hence the biggest is at the bottom. The problem is to move the discs one by one from one peg to another in such a way that no disc should ever be placed on one smaller than itself, and eventually to transfer all the discs from the peg which they rested initially to one of the other pegs (with all the discs in their proper order).

Obviously, the discs may be substituted by cards labelled $1, 2, \cdots, n$. For example, we may use playing-cards (for $n \leq 13$).

For puzzles or problems of this sort, it is a good strategy to experiment with small numbers to see whether there is any pattern that can be exploited to solve the problem. Now if there is only one disc, then the problem is trivial; simply move the disc to any other peg, and we are done. In the case of two discs, move the top (small) disc to any one of the two empty pegs, then the big disc to the remaining empty peg, and finally the small disc on top of the big disc, and our problem is solved.

So far these cases may be too easy, but at least we can observe that

1° The smallest disc can be moved at any time (and to any peg); but moving any disc twice in a row is obviously a waste—Mathematicians always seek out for efficiency. So we shall try not to be wasteful.

2° When we are not moving the smallest disc (say, right after we moved the smallest), there is only one legitimate move we can make—move the smaller of the two discs (that are not the smallest) on top of the bigger disc if no peg is empty. The move is just as trivial if one of the pegs is empty.

Combining these two observations, we realize that the smallest should be moved at every other step (i.e., at every odd-numbered step). And for each of the even-numbered steps, we have one and only one choice. So the move of the smallest disc always determine the next step.

To facilitate our description, we label the discs 1, 2, $\cdots$, n, according to the increasing order of their sizes; and the pegs A, B, C. Furthermore, we designate the cyclic order A-B-C, so that 1+ means moving disc 1 to the next peg in our cyclic order, 2− means moving disc 2 to the next peg in the reversed cyclic order, etc.

Then the 3-disc case can be solved in 7 steps:

$$1+,\ 2-,\ 1+,\ 3+,\ 1+,\ 2-,\ 1+\ .$$

Of course, the solution is not unique. For example,

$$1+,\ 2-,\ 1+,\ 3+,\ 1-,\ 2+,\ 1+,\ 2+,\ 1-\ ;$$

which takes 9 steps. It appears that reversing the moving direction of disc 1 in the cyclic order is the cause of extra steps.

Exercise. Is there an 8-step solution for the 3-disc case?

Let us pause and think before more trials and errors. For the problem to be solvable, it is necessary that we should be able to move the biggest disc at the bottom, and to do so (because the biggest disc can only be moved to an empty peg), we have no choice but to move all other discs to one of the other pegs (with discs in the proper order, of course), so that there is an empty peg to which we can move the biggest disc. But this means that solving the $(n-1)$-disc case is a prerequisite for solving the n-disc case.

On the other hand, if we can move all the discs except the biggest to another peg (i.e., if we can solve the $(n-1)$-disc case), then we can move the biggest to the remaining empty peg, and resume the procedure of solving the $(n-1)$-disc case to move all the $(n-1)$ discs (i.e., except the biggest discs) on to the top of the biggest disc. In this way, we can solve the n-disc case.

In other words, the solvability of the $(n-1)$-disc case is a necessary and sufficient condition for that of the n-disc case. Similarly, the solvability of the $(n-2)$-disc case is a necessary and sufficient condition of the solvability of the $(n-1)$-disc case, etc.

Repeating this reasoning, we eventually arrive at the 1-disc case. But we know that the 1-disc case is solvable; in fact, trivial. Hence the Tower of Hanoi can be solved regardless of the number of the discs.

Note carefully, this is precisely the idea in mathematical induction. Suppose we have a sequence of propositions $P(n)$ that depend on positive integers n, and

(A) $P(1)$ is true (solvable);

(B) "$P(n)$ is true (solvable)" implies "$P(n+1)$ is true (solvable)".

Then propositions $P(n)$ are true (solvable) for all positive integers n.

We saw in the 3-disc case that reversing the moving direction of disc 1 in the cyclic order was not an efficient way to solve the problem. So if we decide that disc 1 should

The Tower of Hanoi

only be moved in the direction of the cyclic order, then from observation 2° above, all the moves are completely determined, and that settles the problem.

Let x_n be the minimum number of steps needed to solve the n-disc case. Then our discussion above tells us that we have

$$x_{n+1} = 2x_n + 1 \quad (n \geq 1).$$

From this relation (and $x_1 = 1$), we can determine the value of x_n either by mathematical induction, or by

$$\begin{aligned} x_{n+1} &= 2x_n + 1 = 2(2x_{n-1} + 1) + 1 = 2^2 x_{n-1} + (2+1) \\ &= 2^2 (2x_{n-2} + 1) + (2+1) = 2^3 x_{n-2} + (2^2 + 2 + 1) \\ &= \cdots \\ &= 2^n x_1 + (2^{n-1} + 2^{n-2} + \cdots + 2 + 1) \\ &= 2^n + 2^{n-1} + 2^{n-2} + \cdots + 2 + 1 = 2^{n+1} - 1. \end{aligned}$$

Exercise. (New Year Puzzle 2012) Our Tower of Hanoi consists of n discs of different sizes (labelled 1, 2, ..., n, according to their increasing sizes), and three pegs A, B, C, fastened to a stand. Initially, the discs are all on peg A with the smallest (Disc 1) at the top, and the sizes of the successive discs increase as we descend. Our mission is to transfer all the discs from peg A to another peg under the rules that

(i) We can move only one disc at each step from one peg to another;

(ii) No disc should ever be placed on one smaller than itself;

(iii) At the odd-numbered steps, Disc 1 is moved to the next peg in the *cyclic* order A-B-C, but at the even-numbered steps, Disc 1 is not allowed to be touched. (This determines all the steps.)

Questions

(a) Show that all the odd-numbered discs are always moved in the cyclic order, while all the even-numbered discs are always moved in the reversed cyclic order.

(b) What is the minimum number of discs so that we need at least 2012 steps to accomplish the mission?

(c) Describe the 2012-th step. In other words, determine which disc should be moved (at the 2012-th step)? And from which peg to which peg?

(d) Find the formula for the minimum number y_n of steps needed to move n discs from peg A to peg B (for n even).

Chapter 9
Ladder Lotteries

In this chapter, we discuss an intriguing lottery known as Amidakuji, in Japanese.[1] ("Amida" is a word in Buddhism, and "Kuji" means lottery.) Let us start with an example. Suppose we want to award books (labelled 1, 2, $\cdots$, 7) to Amy, Dorothy, Heather, Kira, Rosemary, Sierra, and Veronica.

1. We draw 7 vertical lines (posts), assign a book to the bottom of each post, and cover this bottom part. (Figure AMD/Left).

2. Each person selects one of the posts and also draws any number of horizontal segments (rungs) between any pair of neighboring posts as she pleases. To avoid ambiguity, no two rungs are allowed to touch each other.

3. When it is time to award the prizes, we open the covered part at the bottom (which may even hide some rungs constructed before covering up), and let each person slide down from the top of her post to receive the prize at the bottom. She must cross every rung she reaches, then continue sliding down. (Figure AMD/Right)

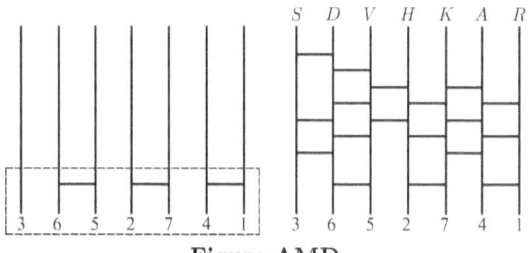

Figure AMD

---

[1] I learned about Amidakuji more than half a century ago from a Japanese book, *Mathematical Puzzles and Games* by MORIMOTO Seigo (Maki Publisher, Tokyo, 1948).

In our case, Veronica and Kira receive books 1 and 2, respectively. The reader is encouraged to find out which books the remaining persons received.

Exercises.

1. Is it possible for two persons to share the same prize? Or, a person to get two prizes?

2. In our example, we constructed 18 rungs. But the same result can be achieved by fewer rungs. How many rungs can be removed and still maintain the same result?

3. From LEAP to PEAL, it can be done with 5 rungs. (Figure AMD1) Is it possible to obtain this same result by constructing only 4 rungs? 6 rungs? 7 rungs?

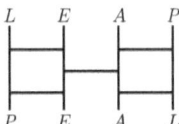

Figure AMD1

Mathematically, Amidakuji are permutations related to sorting algorithms in computer science, among others. Consider the following puzzle.

Place rungs between the posts so that the numbers at the top connect to the numbers at the bottom. (Figure AMD2)

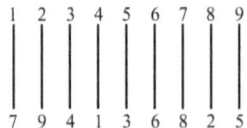

Figure AMD2

Obviously, there must be infinitely many solutions, if any. (Why?) Naturally, we are interested in a solution that needs as few rungs as possible. We shall call such a solution a "minimal ladder".

To solve this kind of problems, a natural approach is to take a close look at how each one moves. We notice that each section of a post is slid down once and only once, while each rung is crossed twice, once in each direction. Moreover, each rung (between a pair of neighboring posts) exchanges the persons who slide down these posts. We keep track of which person crossed which rung by writing down the initial of the person at the end of the rung just before she slides down again. (Figure AMD3/Left)

Ladder Lotteries

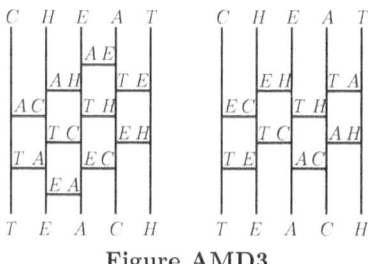

Figure AMD3

Naturally, rung AE exchanges the posts of E and A. Similarly for others. In our example, we have 10 rungs:

$$AC, TA, AH, TC, EA, AE, TH, EC, TE, EH.$$

Of these rungs, two of them, EA and AE, cancel the effects of each other, so removing this pair of rungs does not affect the final result at all. (This is a wasteful, redundant pair of rungs.) (Figure AMD3/Right)

Note that the final state $TEACH$ has 10 ordered pairs:

$$TE, TA, TC, TH, EA, EC, EH, AC, AH, CH.$$

Of course the initial state $CHEAT$ also has 10 ordered pairs:

$$CH, CE, CA, CT, HE, HA, HT, EA, ET, AT.$$

Comparing these two sets of 10 ordered pairs, we notice that EA and CH are common to both sets, while each of the remaining 8 ordered pairs appears in reverse order in the other set, and these 8 ordered pairs (in the final state) are precisely the 8 ordered pairs that appear on the rungs (after removing the redundant pair)!

Is this by accident? Absolutely not. Starting from the initial state $CHEAT$ to obtain the final state $TEACH$, we must move T from the right end to the left end, and to do so, T must exchange the order with A, E, C, and H, one by one. Hence we do need rungs TA, TE, TC, and TH. Similarly, for E to move ahead of C and H, we need rungs EC and EH; etc.

Note that this observation also explains why rung AE is redundant. The order of E and A in the initial state is the same as in the final state; hence constructing rung AE to exchange their order is superfluous. (And once AE is constructed, then we must construct another one, EA, to repair the "damage".) Hence we obtain the following.

> *Theorem.* The number of rungs in a minimal ladder is equal to the number of ordered pairs in the initial state that appear in reverse order in the final state.

Thus, given an Amidakuji, we know how to calculate the minimum number of rungs needed to achieve the same result, and which rungs are redundant, if any. However, consider the following. (AMD4/Left)

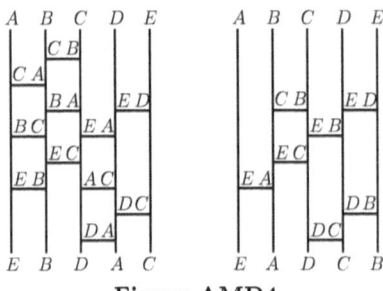

Figure AMD4

There exist two pairs of redundant rungs, CB and BC; CA and AC. Yet removing these two pairs with one stroke, we get a different result. (Figure AMD4/Right) The reason is obvious. These four rungs involve A, B, and C. So it is only natural that (when removing these four rungs) these three ended up in different order. So to obtain the same result, we must remove one pair (of redundant rungs) at a time.

Suppose we remove rungs CB and BC (from Figure AMD4/Left) first. Then because B and C exchange their order twice, and so we obtain the same result. (Figure AMD5/Left)

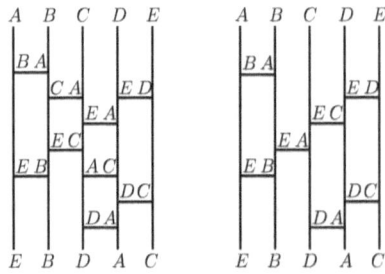

Figure AMD5

Note that rung CA in Figure AMD4/Left becomes BA in Figure AMD5/Left, and BA (in figure AMD4/Left) becomes CA (in Figure AMD5/Left). This is why we should remove one pair at a time. Finally, removing the new pair CA and AC (in Figure AMD5/Left), we obtain a minimal ladder. (Figure AMD5/Right)

Of course from Figure AMD4/Left, we could remove the pair CA and AC first. Then we obtain Figure AMD6/Left.

Ladder Lotteries

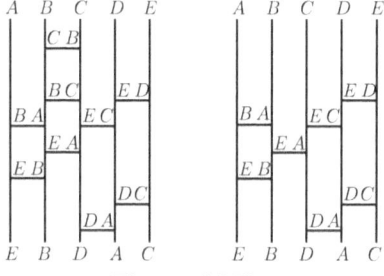

Figure AMD6

Again note that rung BC in Figure AMD4/Left becomes BA in Figure AMD6/Left, and rung BA (in Figure AMD4/Left) becomes BC (in figure AMD6/Left). Now rungs CB and BC obviously cancel the effects of each other, and so removing them gives us a minimal ladder. (Figure AMD6/Right)

In our example, we obtain the same minimal ladder regardless of the order in removing the redundant pairs. However, this is not true in general.

Exercises. (a) How many different minimal ladders are contained in the following ladder?

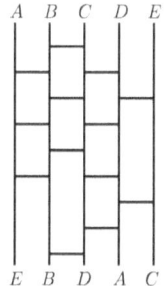

Figure AMD7

(b) What if there is no common element between two pairs of redundant rungs? Can you remove all of them simultaneously?

Theorem. Any Amidakuji contains a minimal ladder.

By now, given an initial state and the final state, it should be clear how to construct a minimal ladder. We present two algorithms by way of an example.

Figure AMD8

We see that 8 ordered pairs,

$$SO,\ SF,\ SE,\ SR,\ OF,\ TE,\ TR,\ ER,$$

in the initial state must be reversed. So we need 8 rungs.

Algorithm 1. (a) We settle one by one from the left end.

1. Move F from the third post to the first by constructing rungs FO and FS. (Figure AMD9/Left) Then $SOFTER$ becomes $\underline{F}SOTER$. (We underline F to indicate that F has been settled.) Note that the remaining 6 ordered pairs (SO, SE, SR, TE, TR, ER) that needed to be reversed do not involve F.

2. Move O from the third post to the second by constructing rung OS. Then $\underline{F}SOTER$ becomes $\underline{FO}STER$. Note again that the remaining 5 ordered pairs (SE, SR, TE, TR, ER) that needed to be reversed do not involve F nor O.

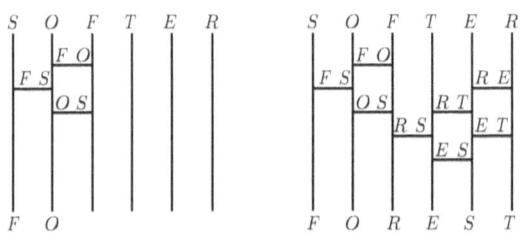

Figure AMD9

3. Move R from the sixth post to the third by constructing rungs RE, RT, and RS. (Figure AMD9/Right) Then $\underline{FO}STER$ becomes $\underline{FOR}STE$. Note that the remaining ordered pairs (SE, TE) that needed to be reversed involve only the unsettled ones, S, T, and E.

4. Move E from the sixth post to the fourth by constructing rungs ET and ES. Then $\underline{FOR}STE$ becomes $\underline{FORE}ST$. And the remaining S and T fall into their proper positions automatically.

(b) There is nothing sacred about the left end. We could start to settle one by one from the right end.

1. Move T from the fourth post to the last by constructing rungs ET and RT. Then $SOFTER$ becomes $SOFER\underline{T}$. (Figure AMD10/Left)

2. Move S from the first post to the fifth by constructing rungs OS, FS, ES, and RS. Then $SOFER\underline{T}$ becomes $OFER\underline{ST}$.

Ladder Lotteries 77

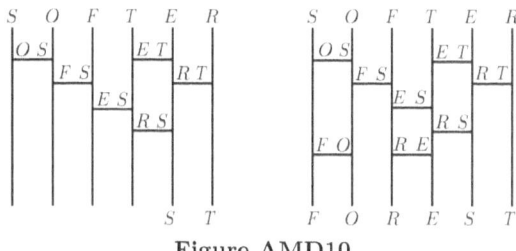

Figure AMD10

3. Move E from the third post to the fourth by constructing rung RE. (Figure AMD10/Right) Then $OFER\underline{ST}$ becomes $OF\underline{REST}$, and R falls into its proper position automatically.

4. Finally, move O from the first post to the second by constructing rung FO. Then $OF\underline{REST}$ becomes $F\underline{OREST}$, and we are done.

(c) We could alternate left and right ends. Thus (Figure AMD11/Left)

$SOFTER \to \underline{F}SOTER \to \underline{F}SOER\underline{T} \to \underline{FO}SER\underline{T} \to \underline{FO}ER\underline{ST} \to \underline{FOREST}$.

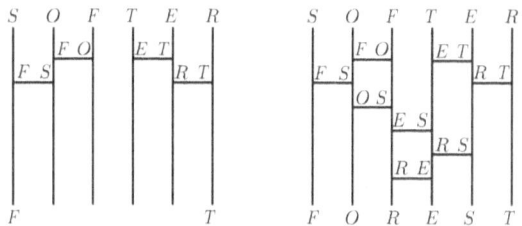

Figure AMD11

Actually, we could settle from the two ends in any manner as long as we keep unsettled part in one consecutive section; i.e., not to cut the unsettled part into two or more separate sections. For example, we could do as follows:

$SOFTER \to \underline{F}SOTER \to \underline{FO}STER \to \underline{FO}SER\underline{T} \to \underline{FO}RSE\underline{T} \to \underline{FOREST}$.

Algorithm 2.

1. Of the eight rungs FO, FS, OS, RE, RS, RT, ES, and ET, that need to be constructed, at this first state we can only construct one of $\{FO, OS\}$, and one of $\{RE, ET\}$. Because if rungs OS and FO are constructed simultaneously, then they would touch each other. So we choose one of them at random, say OS. Similarly, ET and RE cannot be constructed simultaneously, and we choose one of them at random, say RE. Then $SOFTER$ becomes $OSFTRE$.

2. Of the six rungs, FO, FS, RS, RT, ES, and ET, that remain to be constructed, we can construct at this state, FS and RT; and $OSFTRE$ becomes $OFSRTE$. (Figure AMD12/Left)

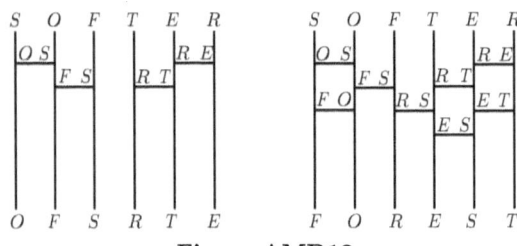

Figure AMD12

3. Of the four rungs, FO, RS, ES, and ET, that remain to be constructed, we can construct at this state, FO, RS, and ET; and $OFSRTE$ becomes $FORSET$. (AMD12/Right)

4. Now it is obvious that all we need is to construct the remaining rung ES; and we are done.

Projects for further investigation.

1. Given the initial and final states, how do we count the number of minimal ladders?

2. What if Amidakuji is pasted on a cylinder so that we can construct rungs between the last and the first posts?

Chapter 10

Round Robin Competitions

Suppose we have 12 table tennis players. Instead of a tournament, during which many pairs of players do not have a chance for a match, the organizer decides on a round robin competition so that every pair of players has a chance to meet. He wants every player to have one match every day so that six matches are played every day. Naturally, it takes eleven days to play all the matches, because each one has 11 opponents. How should he arrange the matches? He is concerned that if not well-planned, some players might find the opponents they needed unavailable in the final days, forcing them to wait and requiring extra days for the competition.

We present three solutions. Let us label players 1 through 12.

Solution 1. Let 11 points that are equally spaced on a circle be labelled 1 through 11, and the center as 12. (Figure RRC/Left) Then connect 1 and 12; 2 and 11; 3 and 10; 4 and 9; 5 and 8; 6 and 7. The pair of players connected by segments are the ones that are matched on the first day. For the second day, third day, and so on, simply rotate the disc (but not the players) by $\frac{2\pi}{11}$, $\frac{4\pi}{11}$, and so on (clockwise, say).

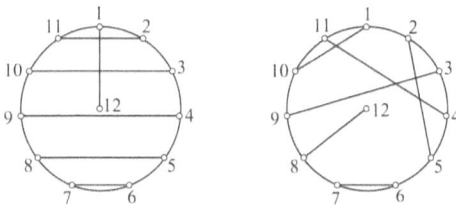

Figure RRC

It is easy to see that all pairs of players are matched in this arrangement as the disc makes a full 2π turn. Because no two chords are of the same length, no pairs of players will meet twice. And because together these five chords exhaust all possible lengths of a chord (with endpoints at these 11 points), every player gets to meet every other player.

Note that as long as five chords are of different lengths, and no two chords share a common endpoint, then the argument works. (Figure RRC/Right)

Remark. I find the solution above in a Japanese book, *Mathematical Puzzles* (Chuou-Kouron Sha, Tokyo, 1975), co-authored by IKENO Shin'ichi, TAKAGI Shigeo, TSUCHIBASHI Sousaku, and NAKAMURA Gisaku. As usual, I did not read the solution until I solved the problem myself. But after solving the problem myself, I rearranged their solution as above for an easy comparison with my solution 2 below.

It is clear that this solution and the following can easily be extended to the case of any even-numbered players. The next solution is my "one-formula" solution.

Solution 2. Except player 12, on the k-th day, players x and y are matched if

$$x + y \equiv k \pmod{11}.$$

Note that for each k ($k = 1, 2, \cdots, 11$), there exist exactly 5 (unordered) pairs of solutions (x, y) for which $x \neq y$; and one solution for which $x = y$. In the latter case, this player x plays against player 12. (I.e, on the k-th day, player 12 plays against player x for whom $2x \equiv k \pmod{11}$.) Hence we have six matches every day. If x is fixed, then for different k, we obtain different y, so no one is going to meet the same player twice; i.e., each player gets to play against every other player in 11 days.

Remark. I was delighted that my solution is so much simpler than the one in the book. So I announced my discovery in our family network. Within hours, my son, Shin-Hong, sent me solution 3 below.

Solution 3. Choose one of the players, say player 12, to be fixed and rotate other players. Thus on the first day, we match the players as follows:

$$\begin{pmatrix} 1 & 2 & 3 & 4 & 5 & 6 \\ 12 & 11 & 10 & 9 & 8 & 7 \end{pmatrix},$$

where two players in the same column play against each other. And from the second day on, we have

$$\begin{pmatrix} 2 & 3 & 4 & 5 & 6 & 7 \\ 12 & 1 & 11 & 10 & 9 & 8 \end{pmatrix}, \quad \begin{pmatrix} 3 & 4 & 5 & 6 & 7 & 8 \\ 12 & 2 & 1 & 11 & 10 & 9 \end{pmatrix}, \quad \text{etc.}$$

It is simple to check that this arrangement satisfies the condition.

Shin-Hong also noticed that his solution can be generalized to odd numbers of players as well. Just eliminate player 12 from the solution above, and have the player who would have been playing player 12 sit out that day, and we have a solution for 11 players. (Naturally, the same can be done for other solutions.)

To be precise, Shin-Hong's original solution lined up players in two increasing rows; thus player 1, who was the fixed player, against player 7 on the first day, and player 12 followed player 6 in the cyclic order. Having spent hours for solution 2 above, I

Round Robin Competitions

quickly rewrote his solution as solution 3 above, again for an easier comparison with mine.

I was basking in glory that my solution lets each player figure out who will be his opponent on any particular day without using pencil and paper when Shin-Hong sent another e-mail saying,

> Another variation of the problem occurs to me: what if the competition director doesn't try to explicitly schedule matches at all? If he just told the players, "match up with whomever you please each day, as long as it's someone you haven't played yet", would that suffice? Or is it possible that some players would find the opponents they needed unavailable in later days, forcing them to wait and requiring extra days for the competition? I suspect the answer to the latter question is "no", but I don't have a rigorous proof yet.

His conjecture is right. If player x has not played against player y, then the converse is also true. So two players are either both available or both unavailable on any given day. Besides, every player has exactly the same number of players he/she has not played yet on any day. Thus on the eleventh day (the final day), every player has no choice but to play against the only player he/she has not played yet. So a haphazard arrangement does work. I should have noticed this at the very beginning. I spent two days for a "non-problem"!

Project for further investigation. What if, instead of a 2-player game, we have a 3-player game such as diamond game (known as Chinese Checkers in the U.S.A), or a 4-player game such as mahjong?

Chapter 11

Egyptian Fractions

The Rhind Papyrus (1650 B.C.), discovered in 1858, is one of the oldest written mathematics. The ancient Egyptians seemed to express fractions as the sums of unit fractions (i.e., fractions with the numerator equals to 1). For example,

$$\frac{2}{29} = \frac{1}{15} + \frac{1}{435} = \frac{1}{16} + \frac{1}{240} + \frac{1}{435} = \frac{1}{15} + \frac{1}{436} + \frac{1}{189660}.$$

Or,

$$\frac{2}{97} = \frac{1}{49} + \frac{1}{4753} = \frac{1}{50} + \frac{1}{2450} + \frac{1}{4753} = \frac{1}{49} + \frac{1}{4754} + \frac{1}{22595762}.$$

These examples suggest that there are many representations for a same fraction.

Theorem. Every positive fractions can be expressed as the sum of finitely many distinct unit fractions. Furthermore, given any positive number ℓ, we may use only those unit fractions whose denominators are greater than ℓ. (Hence there exist infinitely many such expressions.)

The proof of the first assertion by C. Triggs in *Mathematical Quickies* (McGraw-Hill, 1967, Q.154, p.43; Solution p.148) is as follows:
Given $\frac{p}{q}$, express it as the sum of p fractions:

$$\frac{p}{q} = \frac{1}{q} + \frac{1}{q} + \cdots + \frac{1}{q}.$$

Then apply the formula

$$\frac{1}{q} = \frac{1}{q+1} + \frac{1}{q(q+1)}$$

repeatedly. Thus

$$\frac{2}{29} = \frac{1}{29} + \frac{1}{29} = \frac{1}{29} + \left(\frac{1}{30} + \frac{1}{29 \cdot 30}\right) = \frac{1}{29} + \frac{1}{30} + \frac{1}{870}.$$

83

$$\frac{2}{97} = \frac{1}{97} + \frac{1}{97} = \frac{1}{97} + \left(\frac{1}{98} + \frac{1}{97 \cdot 98}\right) = \frac{1}{97} + \frac{1}{98} + \frac{1}{9506}.$$

$$\frac{3}{5} = \frac{1}{5} + \frac{1}{5} + \frac{1}{5} = \frac{1}{5} + \left(\frac{1}{6} + \frac{1}{30}\right) + \left(\frac{1}{6} + \frac{1}{30}\right)$$

$$= \frac{1}{5} + \frac{1}{6} + \frac{1}{30} + \left(\frac{1}{7} + \frac{1}{42}\right) + \left(\frac{1}{31} + \frac{1}{930}\right)$$

$$= \frac{1}{5} + \frac{1}{6} + \frac{1}{7} + \frac{1}{30} + \frac{1}{31} + \frac{1}{42} + \frac{1}{930}.$$

However, this proof has a gap. When the numerator p is large, we are not sure that the same denominators will not keep on appearing. In other words, there is no guarantee that this procedure will end with distinct unit fractions in a finite number of steps.

Here is a correct proof by Fibonacci. Because for any positive number ℓ, we have $\sum_{k=\ell}^{\infty} \frac{1}{k} = \infty$, hence given any positive rational number p/q, there exists a positive integer m $(m \geq \ell)$ such that

$$\sum_{k=\ell}^{m} \frac{1}{k} \leq \frac{p}{q} < \sum_{k=\ell}^{m+1} \frac{1}{k}.$$

If the equality on the left-hand side holds, then we have achieved our purpose. If not, then set

$$\frac{p}{q} - \sum_{k=\ell}^{m} \frac{1}{k} = \frac{p_1}{q_1},$$

where p_1 and q_1 are relatively prime. If $p_1 = 1$, then again we are done. (Note that in this case we have $q_1 > m$.) If $p_1 > 1$, then there must be a positive number n greater than $m+1$ such that

$$\frac{1}{n} < \frac{p_1}{q_1} < \frac{1}{n-1}.$$

It follows that

$$\frac{p_2}{q_2} = \frac{p_1}{q_1} - \frac{1}{n} = \frac{np_1 - q_1}{nq_1},$$

and we have $p_2 = np_1 - q_1 < p_1$, implying that if we choose the largest unit fraction available at each step, then the numerator will be strictly decreasing. But a strictly decreasing sequence of positive integers can not have infinitely many terms, therefore this procedure must end in a finite number of steps.

Naturally, there are many other proofs of this theorem. Our next question is, Do we really need all the positive integers (greater than ℓ)? For example, what if we restrict the denominators to be even integers? This question has an easy answer. Given positive rational number p/q, express it by the method above in two ways:

$$\frac{p}{q} = \frac{1}{m_1} + \frac{1}{m_2} + \cdots + \frac{1}{m_i} = \frac{1}{n_1} + \frac{1}{n_2} + \cdots + \frac{1}{n_j}.$$

Egyptian Fractions

Here we may also require $m_1 < m_2 < \cdots < m_i < n_1 < n_2 < \cdots < n_j$. Then we have
$$\frac{p}{q} = \frac{1}{2m_1} + \frac{1}{2m_2} + \cdots + \frac{1}{2m_i} + \frac{1}{2n_1} + \frac{1}{2n_2} + \cdots + \frac{1}{2n_j},$$
where all the denominators are distinct. This method can be applied to the cases that the denominators are multiples of 3, 4, etc.

But can we restrict the denominators to be odd positive integers? This question is not so simple to answer. For example, to express 1 using odd unit fractions, we need at least 9 terms:

$$\begin{aligned}
1 &= \frac{1}{3} + \frac{1}{5} + \frac{1}{7} + \frac{1}{9} + \frac{1}{11} + \frac{1}{15} + \frac{1}{35} + \frac{1}{45} + \frac{1}{231} \\
&= \frac{1}{3} + \frac{1}{5} + \frac{1}{7} + \frac{1}{9} + \frac{1}{11} + \frac{1}{15} + \frac{1}{21} + \frac{1}{135} + \frac{1}{10395} \\
&= \frac{1}{3} + \frac{1}{5} + \frac{1}{7} + \frac{1}{9} + \frac{1}{11} + \frac{1}{15} + \frac{1}{21} + \frac{1}{165} + \frac{1}{693} \\
&= \frac{1}{3} + \frac{1}{5} + \frac{1}{7} + \frac{1}{9} + \frac{1}{11} + \frac{1}{15} + \frac{1}{21} + \frac{1}{231} + \frac{1}{315} \\
&= \frac{1}{3} + \frac{1}{5} + \frac{1}{7} + \frac{1}{9} + \frac{1}{11} + \frac{1}{15} + \frac{1}{33} + \frac{1}{45} + \frac{1}{385}.
\end{aligned}$$

If we are willing to have two more terms, then we can decrease the largest denominator to 105:
$$1 = \frac{1}{3} + \frac{1}{5} + \frac{1}{7} + \frac{1}{9} + \frac{1}{11} + \frac{1}{33} + \frac{1}{35} + \frac{1}{45} + \frac{1}{55} + \frac{1}{77} + \frac{1}{105}.$$
We also have
$$1 = \frac{1}{3} + \frac{1}{5} + \frac{1}{7} + \frac{1}{9} + \frac{1}{15} + \frac{1}{21} + \frac{1}{27} + \frac{1}{35} + \frac{1}{63} + \frac{1}{105} + \frac{1}{135}, \text{ etc.}$$
If we use Exercise 4 below, then we know there are infinitely many such expressions. However, note that the number of terms is always odd. (Why?)

There are two difficulties when we require that the denominators must be odd. First, the denominator q of a given rational number p/q (in reduced form) must be odd. (Why?) Next, imitating the proof of the theorem above, if we choose the largest unit fraction available at each step, then the numerators do not always decrease monotonically. For example, consider $2/7$:

$$\frac{1}{5} < \frac{2}{7} < \frac{1}{3}, \quad \frac{2}{7} - \frac{1}{5} = \frac{3}{35};$$
$$\frac{1}{13} < \frac{3}{35} < \frac{1}{11}, \quad \frac{3}{35} - \frac{1}{13} = \frac{4}{455};$$
$$\frac{1}{115} < \frac{4}{455} < \frac{1}{113}, \quad \frac{4}{455} - \frac{1}{115} = \frac{1}{10465}.$$

Therefore,
$$\frac{2}{7} = \frac{1}{5} + \frac{1}{13} + \frac{1}{115} + \frac{1}{10465}.$$

Note that the sequence of numerators: $\{2,3,4,1\}$ is not monotone decreasing. And we do not know whether by choosing the largest unit odd fraction at each step will end in finite number of steps.

However, R. Breusch[1] and B.M. Stewart[2] proved independently that every rational number with odd denominator can be expressed as the sum of finitely many distinct odd unit fractions.

Egyptian fractions suggest many interesting problems.

Exercises.

1. Find all the pairs of positive integers a and b $(a < b)$ such that
$$\frac{3}{10} = \frac{1}{a} + \frac{1}{b}.$$

2. Show that for every odd positive integer n, there exists a pair of positive integers a and b $(a < b)$ such that
$$\frac{2}{n} = \frac{1}{a} + \frac{1}{b}.$$

3. Given positive integers m and n, show that
$$\sum_{k=m}^{m+n} \frac{1}{k} = \frac{1}{m} + \frac{1}{m+1} + \cdots + \frac{1}{m+n}$$
is never an integer. What if the harmonic series is replaced by
$$\sum_{k=0}^{n} \frac{1}{a+kd} = \frac{1}{a} + \frac{1}{a+d} + \cdots + \frac{1}{a+nd}$$
where $a \geq 2$, $d \geq 1$?

4. Given an arbitrary odd integer $n \geq 3$, show that there exist three distinct odd positive integers p, q, r such that
$$\frac{1}{n} = \frac{1}{p} + \frac{1}{q} + \frac{1}{r}.$$

5. Find a non-empty finite subset S of positive integers such that
   (a) No element of S is isolated; i.e., if n is in S, then either $(n-1)$ or $(n+1)$ must be in S;
   (b) $\sum_{n \in S} \frac{1}{n}$ is an integer.

---
[1] Solutions to Problem E4512, *American Mathematical Monthly*, 61 (1954), 200-201.
[2] *Theory of Numbers*, (Macmillan 1964), 198-207.

Egyptian Fractions

Here is an open problem that I raised more than thirty years ago.[3]

An Open Problem. Is it true that whenever the set of positive integers are partitioned into finite number of subsets, one of the subsets has the property that an arbitrary positive rational number can be expressed as the sum of reciprocals of a finite number of distinct elements of the subset? (Note that here it must be possible to choose the subset independent of the rational number.) If this is not possible, then given any rational number, can we always choose one of the subsets with this property? (Now the subset depends on the given rational number.)

---

[3] Problems and Solutions, *Sugaku*, Mathematical Society of Japan, Vol. 31, No. 4 (October 1979), 376; Egyptian Fractions, *Mathmedia*, Academia Sinica, Taipei 15, (1980), 8 - 12

Chapter 12

The Ptolemy Theorem

12.1 The Ptolemy Theorem

The Ptolemy theorem is one of the gems in Greek mathematics, and it has interesting applications.

Theorem 1. In an arbitrary convex quadrangle $ABCD$ (Figure PTL),

$$\overline{AB} \cdot \overline{CD} + \overline{BC} \cdot \overline{AD} \geq \overline{AC} \cdot \overline{BD}.$$

Equality holds if and only if quadrangle $ABCD$ is cyclic.

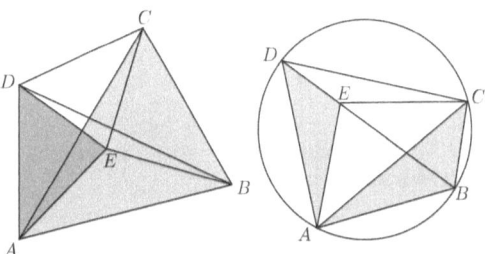

Figure PTL

Proof. Let E be the point satisfying $\triangle AED \sim \triangle ABC$; i.e., choose point E such that

$$\angle EAD = \angle BAC \quad \text{and} \quad \angle EDA = \angle BCA.$$

Then, we have

$$\frac{\overline{AD}}{\overline{ED}} = \frac{\overline{AC}}{\overline{BC}}, \quad \overline{AD} \cdot \overline{BC} = \overline{AC} \cdot \overline{ED}.$$

89

Furthermore, because

$$\angle BAE = \angle CAD \quad \text{and} \quad \frac{\overline{AB}}{\overline{AE}} = \frac{\overline{AC}}{\overline{AD}},$$

we obtain $\triangle ABE \sim \triangle ACD$, implying

$$\frac{\overline{AB}}{\overline{BE}} = \frac{\overline{AC}}{\overline{CD}}; \text{ i.e., } \overline{AB} \cdot \overline{CD} = \overline{AC} \cdot \overline{BE}.$$

Therefore,

$$\overline{AB} \cdot \overline{CD} + \overline{AD} \cdot \overline{BC} = \overline{AC}(\overline{BE} + \overline{ED}) \geq \overline{AC} \cdot \overline{BD}.$$

Equality holds if and only if point E is on BD, which occurs if and only if $\angle ADB = \angle ACB$; i.e., if and only if quadrangle $ABCD$ is cyclic, and our proof of the Ptolemy theorem is complete. □

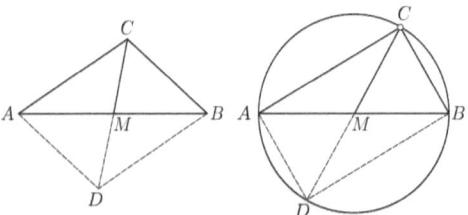

Figure PTL1

Note that for an arbitrary triangle ABC, we can find point D such that $ADBC$ is a parallelogram (Figure PTL1), and the theorem above gives

$$\overline{BC}^2 + \overline{CA}^2 \geq 4\overline{AM} \cdot \overline{CM},$$

where M is the midpoint of side AB. And parallelogram $ADBC$ is cyclic if and only if, in the original triangle ABC, angle C is a right angle, and in this case the theorem gives the Pythagorean theorem.

Exercise. Given a cyclic quadrangle, express the lengths of the two diagonals in terms of the lengths of the four sides.

12.2 Applications

We now use the theorem to derive the addition formulas for the sine and cosine functions.

$$\begin{aligned} \sin(x+y) &= \sin x \cdot \cos y + \cos x \cdot \sin y; \\ \cos(x+y) &= \cos x \cdot \cos y - \sin x \cdot \sin y. \end{aligned}$$

The Ptolemy Theorem

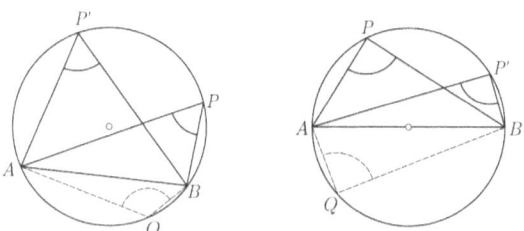

Figure PTL2

Recall, in a circle with diameter 1, the length of its chord is the value of the sine function of an angle it subtends (regardless of whether the vertex of the angle is on the major or minor arc—two angles are supplementary, hence the values of the sine function are equal); in particular, the subtended angle is a right angle if and only if the chord is a diameter. (Figure PTL2)

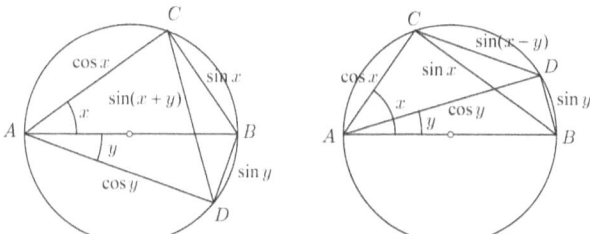

Figure PTL3

In Figure PTL3/Left, let AB be a diameter, and without loss of generality, we may assume $\overline{AB} = 1$. Choose points C and D such that $\angle BAC = x$, $\angle BAD = y$. Then, by our remark above, $\overline{BC} = \sin x$, and because $\angle ACB$ is a right angle, $\overline{AC} = \cos x$. Similarly, $\overline{BD} = \sin y$ and $\overline{AD} = \cos y$. Furthermore, because $\angle CAD = x + y$, we have $\overline{CD} = \sin(x+y)$. Hence, by the Ptolemy theorem, we have

$$1 \cdot \sin(x+y) = \sin x \cdot \cos y + \cos x \cdot \sin y,$$

which is the addition formula for the sine function we want to prove.

Exercise. Prove
$$\sin(x-y) = \sin x \cdot \cos y - \cos x \cdot \sin y$$
using Figure PTL3/Right.

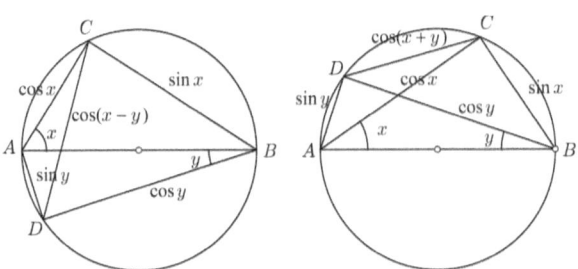

Figure PTL4

Next in Figure PTL4/Left, let AB be a diameter whose length is 1. Choose points C and D such that $\angle BAC = x$, $\angle ABD = y$. Then, we have

$$\overline{BC} = \sin x, \quad \overline{AC} = \cos x, \quad \overline{AD} = \sin y, \quad \overline{BD} = \cos y,$$

and by the Ptolemy theorem, we obtain

$$\overline{AB} \cdot \overline{CD} = \cos x \cdot \cos y + \sin x \cdot \sin y.$$

But $\overline{AB} = 1$, and so it remains to show $\overline{CD} = \cos(x - y)$. Note that $\angle ADB$ is a right angle, hence we have

$$\begin{aligned} \angle CAD &= \angle CAB + \angle BAD = \angle CAB + \left(\frac{\pi}{2} - \angle ABD\right) \\ &= \frac{\pi}{2} + (\angle BAC - \angle ABD) = \frac{\pi}{2} + (x - y). \end{aligned}$$

It follows that
$$\overline{CD} = \sin\left(\frac{\pi}{2} + (x - y)\right) = \cos(x - y),$$

as desired.

Exercise. Prove
$$\cos(x + y) = \cos x \cdot \cos y - \sin x \cdot \sin y,$$
using Figure PTL4/Right.

12.3 A Generalization of the Ptolemy Theorem

Finally, we give an easy proof of a generalization of the Ptolemy theorem using complex numbers.

Theorem 2. For any quadrangle $ABCD$, we have

$$(xy)^2 = (ac)^2 + (bd)^2 - 2abcd \cdot \cos(\angle A + \angle C),$$

where $x = \overline{AC}$ and $y = \overline{BD}$ are the lengths of the two diagonals, and $a = \overline{AB}$, $b = \overline{BC}$, $c = \overline{CD}$, $d = \overline{DA}$ are the lengths of the four sides.

The Ptolemy Theorem

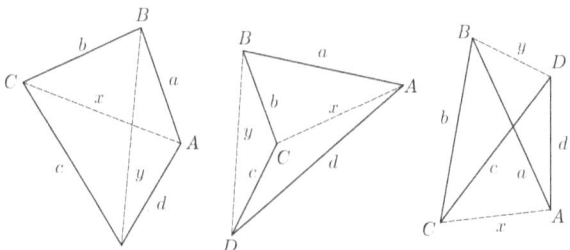

Figure PTM

Proof. [1] Given four arbitrary points $A(\alpha)$, $B(\beta)$, $C(\gamma)$, $D(\delta)$, in the complex plane, it is simple to verify the identity:

$$(\beta - \alpha)(\delta - \gamma) + (\gamma - \beta)(\delta - \alpha) = (\alpha - \gamma)(\beta - \delta).$$

Multiply this identity by its complex conjugate. On the right-hand side, we obtain $(xy)^2$. Of the four terms on the left-hand side, two of them are $(ac)^2$ and $(bd)^2$, respectively.

The remaining two cross terms are complex conjugate of each other, and their (common) absolute value is

$$|(\beta - \alpha)(\delta - \gamma)(\bar{\gamma} - \bar{\beta})(\bar{\delta} - \bar{\alpha})| = abcd,$$

hence the sum of the two cross terms must be $2abcd \cdot \cos\varphi$, where φ is the argument of $(\beta - \alpha)(\delta - \gamma)(\bar{\gamma} - \bar{\beta})(\bar{\delta} - \bar{\alpha})$. So it remains to determine the argument φ. But this is easy. Let θ be the argument of $\overrightarrow{AB}$. Then we have

vector	argument
$\overrightarrow{BC}$	$\theta + (\pi - \angle B)$
$\overrightarrow{CD}$	$\theta + (\pi - \angle B) + (\pi - \angle C) \equiv \theta - \angle B - \angle C \pmod{2\pi}$
$\overrightarrow{AD}$	$\theta + \angle A$

Note that we consider all angles to be oriented; i.e., positive if it is counterclockwise, but negative if clockwise. It follows that

$$\varphi = \theta - [\theta + (\pi - \angle B)] + [\theta - \angle B - \angle C] - [\theta + \angle A] = -\pi - \angle A - \angle C;$$

hence the sum of the two cross terms is

$$2abcd \cdot \cos(-\pi - \angle A - \angle C) = -2abcd \cdot \cos(\angle A + \angle C),$$

and our proof is complete. □

---

[1] I am grateful to Professor Richard Askey (University of Wisconsin, Madison) for suggesting whether this generalization of the Ptolemy theorem can be proved using complex numbers.

Note that the inequalities

$$|ac - bd| \leq xy \leq ac + bd$$

follows immediately from the theorem. The equality on the right-hand side holds if and only if $\cos(\angle A + \angle C) = -1$; i.e., if and only if $\angle A + \angle C = \pi$, and this is true if and only if points A, B, C, D are cocyclic in this order. Hence the theorem includes that of Ptolemy as a particular case. Note also that the equality on the left-hand side forces the quadrangle to degenerate to a line segment.

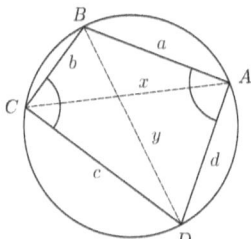

Figure PTM2

Exercise. Multiply the identity below by its complex conjugate and imitate our proof above. What do you get?

$$(z_2 - z_1) + (z_3 - z_2) = z_3 - z_1.$$

Can you find other applications of this method?

Chapter 13

Convexity

13.1 Introduction

The following problem comprises Chapter 16 of the excellent book, *Enjoyment of Mathematics*, by H. Rademacher and O. Toeplitz:

Consider a finite set S of n distinct points $P_1, P_2, \cdots, P_n$, in a plane. Let d be the maximum distance between a pair of points in S; i.e.,

$$d = \max\left\{\overline{P_iP_j}\,;\, 1 \leq i \leq n,\, 1 \leq j \leq n\right\}.$$

With only a finite number of points in S, clearly, there must be a maximum distance. This maximum distance between a pair of points in S is called the *span* of S.

A circle with the property that all the points of S are either inside or on the circle itself is called an *enclosing circle* of set S. (In this chapter, it is convenient to consider a circle as the union of its interior and the boundary; i.e., the so-called closed disc.)

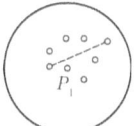

Figure CVX

Because S is a finite set of points in a plane, enclosing circles do exist. Simply take any point, say P_1, in S, and draw the circle centered at P_1 with radius d, the span of S. This circle, denoted as $C(P_1; d)$, is obviously an enclosing circle of S. (Figure CVX)

Given d and n, our problem is to find the minimum radius r (in terms of d and n) that guarantees the existence of an enclosing circle (with radius r) of any set S with n points and span d. Our example, $C(P_1; d)$, shows that $r \leq d$ regardless of the number of points in S.

As usual, our investigation starts from the cases that the number n of the points in set S is small. If $n = 2$; i.e., $S = \{P_1, P_2\}$, then obviously the smallest enclosing circle of set S is the one with diameter $\overline{P_1 P_2}$. Hence $r = \frac{d}{2}$.

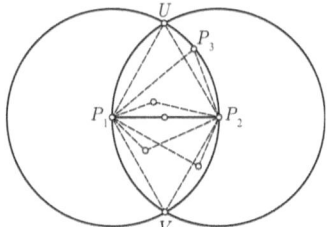

Figure CVX1

Suppose $n = 3$. Let $S = \{P_1, P_2, P_3\}$. We may assume that $\overline{P_1 P_2} = d$, the span of S. Then P_3 must be inside circle $C(P_1\,;d)$ as well as $C(P_2\,;d)$. So P_3 must be in the part of the plane that is inside both circles.

Let U and V be the intersections of the two circles. By symmetry, we may assume, without loss of generality, that P_3 is on the same side as U of line $P_1 P_2$.

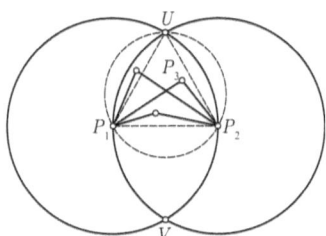

Figure CVX2

Because this region (Figure CVX2) that is half of the lens-shaped region UP_1VP_2 can be enclosed by the circumcircle of $\triangle UP_1P_2$ (whose radius is $\sqrt{3}d/3$), we see that $r = \frac{\sqrt{3}}{3}d$ will do for $n = 3$. On the other hand, by considering the case that point P_3 is at U, we see that this radius can not be improved.

Summing up, we have shown that, for any three points, if the distance between any pair of points is not more than d, then there exists an enclosing circle of radius $\frac{\sqrt{3}}{3}d$, and this is the best possible.

13.2 The Theorems of Jung and Helly

Finally, we consider the general case $n \geq 4$.

Convexity

Theorem. (H.W.E. Jung) For any finite set S of points in a plane, there exists an enclosing circle of S with radius

$$r = \frac{\sqrt{3}}{3}d \quad \text{(where } d \text{ is the span of set } S\text{).}$$

Proof. Our proof is by mathematical induction.

Let J_n be the proposition that, for any n-point set S with span d, there exists an enclosing circle of S with radius $r = \frac{\sqrt{3}}{3}d$.

Suppose $n = 4$. Let $S = \{P_1, P_2, P_3, P_4\}$ be a 4-point set. Because of our work in the previous section, we know any 3-point set has an enclosing circle with radius $r = \sqrt{3}d/3$. Let Q_1, Q_2, Q_3, Q_4, be the centers of enclosing circles of radius $r = \sqrt{3}d/3$ of 3-point sets,

$$\{P_2, P_3, P_4\}, \{P_1, P_3, P_4\}, \{P_1, P_2, P_4\}, \{P_1, P_2, P_3\},$$

respectively. We have two possible cases:

(i) One of the four points, Q_1, Q_2, Q_3, Q_4, is enclosed by the triangle with other three points as its vertices. (The case all four points are collinear can be considered as a degenerated case of this one.) (Figure CVX3/Left)

(ii) None of the four points is enclosed by the triangle with other three points as its vertices; hence these four points are the vertices of a convex quadrangle. (Figure CVX3/Right)

A set is *convex* if it contains the entire segment whenever it contains the two endpoints. Thus circles and triangles are convex, but neither a cross nor a star is convex.

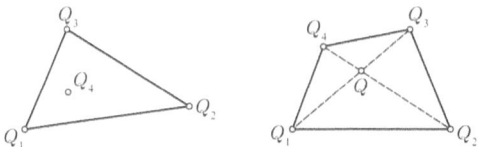

Figure CVX3

Case (i). Without loss of generality, we may assume that point Q_4 is enclosed by $\triangle Q_1 Q_2 Q_3$. (CVX3/Left) We claim: $C(Q_4; r)$ is an enclosing circle of our 4-point set $S = \{P_1, P_2, P_3, P_4\}$. By definition, $C(Q_4; r)$ encloses points P_1, P_2, P_3. So it remains to show that $C(Q_4; r)$ also encloses P_4. Now P_4 is enclosed by all three circles $C(Q_1; r), C(Q_2; r), C(Q_3; r)$, which means that $C(P_4; r)$ encloses three centers Q_1, Q_2, Q_3; i.e., $C(P_4; r)$ encloses $\triangle Q_1 Q_2 Q_3$. (Why?) But Q_4 is enclosed by $\triangle Q_1 Q_2 Q_3$. Therefore, $C(P_4; r)$ encloses Q_4. Equivalently, $C(Q_4; r)$ encloses P_4, which is what we want to show.

Case (ii). Suppose $Q_1Q_2Q_3Q_4$ is a convex quadrangle. (CVX3/Right) Let Q be the intersection of diagonals Q_1Q_3 and Q_2Q_4. Because each of the circles $C(Q_1; r)$ and $C(Q_3; r)$ encloses points P_2 and P_4, each of the circles $C(P_2; r)$ and $C(P_4; r)$ encloses points Q_1 and Q_3, implying that the entire diagonal Q_1Q_3 is enclosed by each of these two circles, $C(P_2; r)$ and $C(P_4; r)$. Similarly, the entire diagonal Q_2Q_4 is enclosed by each of circles $C(P_1; r)$ and $C(P_3; r)$. Thus the intersection Q is enclosed by each of these four circles, implying that circle $C(Q; r)$ encloses the 4-point set $S = \{P_1, P_2, P_3, P_4\}$.

We have shown that "J_3 implies J_4". "J_4 implies J_5" can be done similarly; in fact, even easier (and so we leave it as an exercise for the reader). In general, we have "J_n implies J_{n+1} for $n \geq 3$". And so the theorem is true for all positive integers by mathematical induction. □

Exercise. The following problem appeared in the New Mexico Mathematics Contest (November 2002).[1]

(a) What is the radius of the smallest circle that will enclose any triangle of unit perimeter?

(b) What is the radius of the smallest circle that will enclose any planar 2002-gon (possibly self-intersecting) of unit perimeter?

By exactly the same reasoning, we can prove the following general result.

Theorem (E. Helly) If each three of a collection of n convex sets in a plane have a point in common, then there must be a point that is common to all n of the sets.

Exercise.

(c) What if in the Helly theorem, the condition that "each three" (of a collection of convex sets) is replaced by "each two"?

(d) Can the number of convex sets in the Helly theorem be infinite? Justify your assertion. If the answer is "No", then what additional condition must be imposed on these convex sets?

(e) Can the Helly theorem be generalized to the 3-dimensional case?

---

[1] See the marvelous book by Ross Honsberger, *Mathematical Delights* (MAA 2004), pp. 167 - 176.

Chapter 14

The Seven Bridges of Königberg

14.1 Unicursal Figures

In the 18th century, in the town of Königberg, where famous philosopher Immanel Kant (1724 - 1814) lived his entire life, there were seven bridges over river Pregel as in Figure SBK. A question was raised whether it was possible to take a walk (starting from anywhere) so as to cross every bridge once and only once.

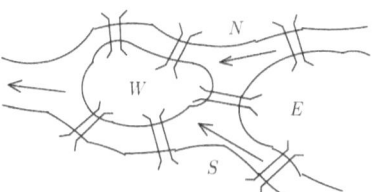

Figure SBK

Certainly, we can label these bridges as 1, 2, 3, 4, 5, 6, 7, and interpret 7-digit number, say 1634527, as crossing bridges 1, 6, 3, 4, 5, 2, 7, in this order. Thus, with patience and stamina, we can determine whether such a walk is possible or not. However, such a solution is not only dull, but with more bridges, we could easily become dizzy. Amazingly, Leonhard Euler (1707 - 1783) is said to solve the problem not only instantly, but also for the general case of any number of islands connected in any way by bridges.

Clearly, the answer will not be affected if we pretend the islands shrink to points and bridges lengthen. Thus we obtain Figure SBK1. Hence the problem becomes whether a given figure can be drawn by one stroke; i.e., whether a figure can be described by a point moving in such a way as to traverse every line once and only once. Suppose a figure can be drawn by one stroke. Then there are two possibilities: Either the starting point and the ending point are the same or different.

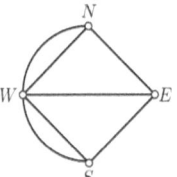

Figure SBK1

Let the starting point be A. If there are more than one paths that meet at A, then first we exit A along one of the paths that meet at A, and return to A along other path, then exit A along yet another path. We repeat "enter/exit" until all the paths that meet at A are exhausted. Therefore, the number of paths that meet at A must be odd (i.e., A is an odd vertex) unless A is also the ending point, in which case A is an even vertex.

At the ending point that is not the starting point, we repeat "enter/exit", and finally ends with "enter". So the ending point is also an odd vertex (if it is different from the starting point).

At each of the vertices that is neither the starting point nor the ending point, we simply repeat "enter/exit" until all the paths that meet at the vertex are exhausted, so all such vertices (i.e., those that are neither the starting point nor the ending point) must be even vertices.

We have shown:

> If a connected figure can be drawn in one stroke, then the number of odd vertices must be either 0 or 2.

With this knowledge, we look at the seven bridges of Königberg. (Figure SBK1) We notice that four vertices W, E, N, S are all odd vertices, and so we conclude that there exists no route that passes every bridge once and only once.

Note carefully, we still do not know that given a connected figure with 0 or 2 odd vertices, whether it is always possible to draw the figure in one stroke. Clearly, this is a different question. Suppose a certain college admits only the national merit scholarship winners. Does this mean all the national merit scholarship winners are admitted to the college?

First we consider the case that our connected figure does not have any odd vertices. Suppose each edge (path) of the figure is represented by a string. At each vertex, we tie strings pairwise together. Because we are assuming that all the vertices are even, we can tie the strings together in pairs so that there will be no loose end. If this tying produces a single loop, then we are done; i.e., starting at any point, we can move along every edge and come back to the starting point. In fact, we may even choose

The Seven Bridges of Könisberg

any one of the edges that meet at the starting point as the first (or the last, if we wish) leg of the route.

But if this tying produces two or more loops, then certainly there must be two loops that touch each other at some vertex (because we are assuming that our figure is connected). Now untie a pair of strings from each of the loops at this vertex and retie them in any other way. Then these two loops will be combined into one loop. Repeating this process if necessary, eventually all the loops will be combined into a single loop.

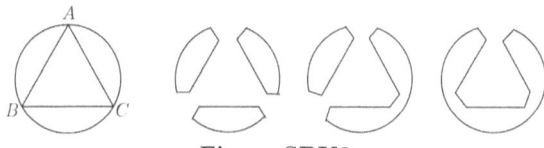

Figure SBK2

For example, consider a triangle inscribed in a circle. (Figure SBK2/Left) At the first tying, we may obtain three loops, but by retying, we can reduce the number of loops one by one, and eventually we obtain a single loop.

It remains to consider the case of a connected figure with two odd vertices, A and Z. In this case, we add an extra edge that connects A and Z. Then we obtain a new connected figure all of whose vertices are even, and so by what we discussed above, this new figure can be drawn by one stroke with A as a starting point and new edge ZA as the last leg of the route. Then the effect of removing this new edge ZA amounts to restoring the original figure, and our route becomes starting from A and ending at Z.

Exercises.

(a) Prove that in any figure, the number of odd vertices is even.
    Hint. Consider each of the edges as a string, and tie the strings pairwise that meet at a vertex, so that each odd vertex corresponds to one loose end of a string.

(b) Prove that a connected figure with $2n$ odd vertices can be drawn in n strokes.

(c) How many ways to draw Figure SBK2/Left in one stroke with A as the starting point? What if a quadrangle is inscribed in a circle (instead of a triangle)?

14.2 Mazes

Some amusement parks have mazes, and many people have ventured inside. But how to make sure that we won't get lost in a maze?

102   Mathemagical Buffet

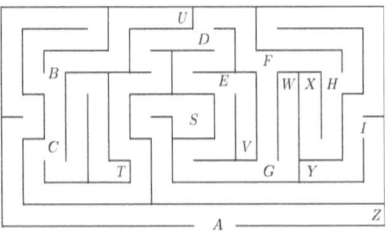

Figure MZS

Here is a brilliant idea. Imagine that the walls of the maze are ditches. And we pump water into the ditch at the entrance on the right of A. If the ditches are connected so that the water eventually flows to the ditch at the exit on the left of A, then we may regard the ditches as a pond, regardless of its shape. And we know if we walk along the boundary of a pond (ditch) so as to always see the pond to our right, then eventually, we will come to any point on the same connected component of the boundary of the pond. In particular, because the entrance is to the right of A, while the exit is to the left, we should be able to come to the exit. (Of course, if we wish, we may walk in such a way so as to always see the pond to our left.)

If we carry out our plan in Figure MZS, then we'll hit dead end Z and be forced to come back to entrance A. But here we are more concerned about our safe return, so making a shortcut is not our first priority.

Note that this rule, "Walk so as to see the wall to your right", must be observed from the beginning. For, if we start to follow the rule only after we are at point F facing G, then we will be making a round ($FGEVED$ and back to F) over and over, and we will never get out of the maze because there is an "island" in the area surrounded by $DEVGF$ (where the ditch there gets no water); i.e., the boundary of the island is not connected to the boundary of the pond that A is on.

On the other hand, following the rule from the beginning does not guarantee that we get to see every chamber in the maze. For example, suppose your sweet-heart is waiting for you at point V in Figure MZS, while you are at entrance A. Then you will not be able to meet your sweet-heart by following the rule.

So, is there a way that guarantees us that we get to see every chamber in the maze without its map? We know every connected figure can be traversed by one stroke if every edge is traversed twice (because now every vertex is even). Of course, to walk twice over every path in a maze is not the fastest way of arriving at the center, but that is not our main concern now.

So consider the dead ends and also the places in a maze where two or more paths meet as vertices, and paths between vertices as edges. Now for our maze (Figure MZS), starting from entrance A, if we go to the right, then we'll hit the dead end at Z, and we'll be forced to come back to A. From A, if we go to the left, we'll come to B. From B, if we go to the left, we'll hit the dead end at U; if we go to the right, then we'll come to C. Continuing in this way, we obtain an abbreviated map (Figure

The Seven Bridges of Königberg

MZS1).

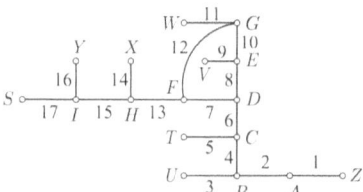

Figure MZS1

As we can see from this map, there is a loop $DEGF$. How do we avoid the trap of going around the loop, yet be able to walk every path once and only once in each direction? Here are rules that guarantee just that, but we need markers or pegs to keep track of paths and directions that we traversed (say, by placing a marker on the floor at the entrance of each path we are entering):

After traversing path PQ to vertex Q,

1. if you have not been to Q before, then proceed (if possible) along any other path QR; but if Q is a dead end, then naturally, you must go back to P.
2. if you have been to Q before, and
   (a) if there is a path QR that has not been traversed (in either direction), then go to R.
   (b) if all the paths that meet at Q have been traversed at least once, then
      i. go back to P (if possible);
      ii. if PQ has been traversed in both directions already, then (recalling that no path should be traversed twice in the same direction) choose QR that has not been traversed in the direction away from Q.

Note that if we follow the rules, then we won't be stuck in a place. Because we are traversing each path twice, all the vertices are even, and so at each vertex, we'll simply repeating "enter/exit". Hence there is no chance that we'll end with "enter" unless that vertex is the starting point, in which case it will be a happy return. O.K., it may be a happy return, but we must make sure that we have traversed every path. Even though we are back to the starting point (because the starting point is also a vertex), as long as there is a path that has not been traversed, then by rule 2(b), we must keep on going. Note that there will be no path left without being traversed as long as it is connected (directly or indirectly) to the starting point because of rule 2(b)i.

For example, consider the abbreviated map (Figure MZS1). We label paths as 1, 2, 3, $\cdots$, and use prime (') to denote the second time we traverse that path. Our route could be described as follows:

A-1 Z-1' A-2 B-4 C-6 D-8 E-10 G-12 F-7 D-7' F-13 H-15 I-16 Y-16' I-17
S-17' I-15' H-14 X-14' H-13' F-12' G-11 W-11' G-10' E-9 V-9' E-8' D-6'
C-5 T-5' C-4' B-3 U-3' B-2' A

(Actually, we could describe our route by using letters only or the numerals only, but economy is not our concern here.) Perhaps some explanation might be in order.

• After F-7, we arrive at D for the second time, so we must go back to F according to rule 2(b)i (because both DE and DC have been traversed at least once).

⋆ After D-7' and back to F, we cannot take path 12' (to go to G) because path 13 has not been traversed yet (by rule 2(a)).

⋆ Similarly, after Y-16' and back to I, we cannot take 15', because path 17 has not been traversed yet (by rule 2(a)).

⋆ Similarly, after I-15 and come back to H, then we must choose path 14 (to X) by rule 2(a).

• After H-13' and come back to F, path 7 has been traversed twice already, so we must choose path 12' (by rule 2(b)ii).

⋆ After F-12 and back to G, we must choose path 11 to W (by rule 2(a)).

⋆ Same reason when we are back to E after G-10', we must choose path 9.

⋆ Same reason when we are back to C after D-6', we must choose path 5.

⋆ Same reason when we are back to B after C-4', we must choose path 3.

Exercise. Reconstruct the abbreviated map based only on our route above.

Chapter 15
The Euler Formula

15.1 The Euler Formula

For polygons in a plane, clearly, the numbers of the vertices and the sides are equal whether they are convex or concave. What can we say for solids in a (3-dimensional) space? It is only natural that for solids, in addition to vertices and edges, we should also take faces into account. Let us start with some observations (Figure EFL):

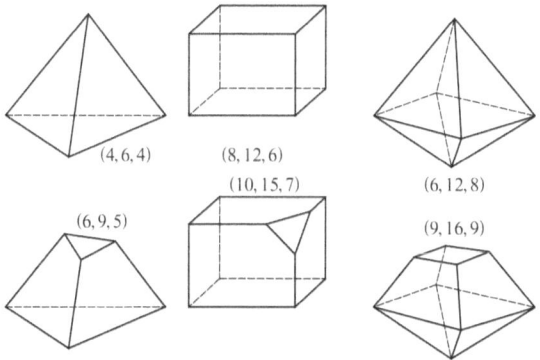

Figure EFL

Three numbers in the parentheses are the numbers of vertices, edges, and faces in each case. We denote these numbers by V, E, and F, respectively. Can you find a relation among these numbers? If not, do not be discouraged. Solids have been studied by Greek geometers intensively, but the beautiful and important formula that is the subject of this chapter was discovered only in 18th century by Leonhard Euler (1707 - 1783).

Here is a hint. Compare the two solids in the left column in Figure EFL. We chop off one corner (a tetrahedron; i.e., a pyramid with a triangular base) from the solid at the top to get the one at the bottom. In the process, one (old) vertex was replaced by 3 new ones (hence the number of vertices has a net gain of $3-1=2$), and we gain 3 new edges and one new face. (Of course 3 of the original edges become shorter, and the three triangular faces become quadrangular, but we are only concerned with the numbers not the length, shape or area.) Exactly the same phenomenon occurs when we cut a corner from a cube. (The middle column in Figure EFL) In the case of solids in the right column in Figure EFL, because the base of the pyramid that we chop off is a quadrangle, the number of vertices increases by $4-1=3$, while the numbers of edges and faces increase by 4 and 1, respectively.

It is now easy to see that, in general, when we chop off a pyramid (with a polygon of m sides as its base), then V, E and F become $V' = V + (m-1)$, $E' = E + m$ and $F' = F + 1$, respectively. Thus by eliminating m from these relations, it is easy to see that
$$V' - E' + F' = V - E + F.$$
That is, $V-E+F$ remains unchanged (in other word, invariant) under the operation of chopping off a pyramid of any shape. So, in all three cases, the number $V-E+F$ for the solid at the top row is the same as that of the corresponding one at the bottom row.

Checking the case of a tetrahedron (a pyramid with a triangular base), we obtain $V-E+F = 4-6+4 = 2$. For a cube, we obtain $V-E+F = 8-12+6 = 2$. How about the one on the right? We have $V-E+F = 6-12+8 = 2$. It seems that 2 appears to be the magic number for any solid, and we have a conjecture: For any (connected) solid,
$$V - E + F = 2.$$
But can we justify this magic formula? Is it true in general? For the sake of simplicity, we call it "solid", but what really concerns us here is the surface of a solid. So imagine that the surface of a solid is a thin rubber with unlimited elasticity, and we puncture a hole in one of the faces so that we can stretch the rest of the faces until they are flat on a plane. (Figure EFL1) Naturally, the edges may become crooked, but the numbers of the vertices and edges remain unchanged. So is the number of faces if we consider the punctured face corresponds to the unbounded region in the plane. (It will not affect our reasoning below, but as long as we are assuming that the elasticity of the rubber is unlimited, we may consider all the edges are straight.)

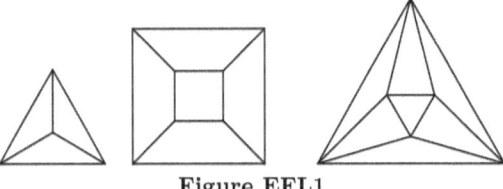

Figure EFL1

Now, I don't know who came up with the following wonderful idea. Consider the

The Euler Formula

figure obtained above as a map of dykes and rice paddies. Rice paddies are dry but the unbounded region is covered by water, and we want to break dykes one by one until every rice paddy gets water. (Figure EFL2) Certainly, we could break all the dykes to water all the rice paddies. But we want to do it in such a way as to break as few dykes as possible. Clearly, then, a dyke that has water on both sides of it need not be broken. And if we break a dyke that has water on just one side, the number of rice paddies that have water increases by one. Recalling that the unbounded region corresponds to one of the faces in the solid, the number of rice paddies is $F - 1$, and so all we need is to break $F - 1$ number of dykes to water all the rice paddies.

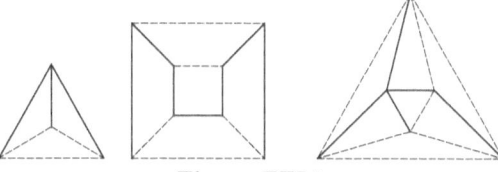

Figure EFL2

So, after breaking $F - 1$ dykes, how many dykes remain standing? To count the number of the remaining dykes, first observe that all the vertices are still connected by the remaining dykes. Because at the beginning, certainly all the vertices are connected, and in the process of breaking the dykes one by one, if there was ever a dyke whose destruction causes some two vertices to be disconnected, then there must be water on both sides of this dyke, hence this dyke should not be destroyed.

Hence any pair of vertices must still be connected by the remaining dykes. Actually, we can make a stronger assertion. That is, given any pair of vertices, not only they are connected, but there exists exactly one route that connects them. For, if there exist two different routes that connect a pair of vertices, then these two routes form a loop, and so any rice paddy inside this loop must remain dry, contradicting our assumption that all the rice paddies are watered.

Now if we fix one vertex to be the starting point, then there exists one and only one route that leads to each vertex. And on each route there exists one and only one last dyke (edge) that must be traversed before reaching the vertex. Thus, with a starting vertex fixed, there is a one-to-one correspondence between the set of other $V - 1$ vertices and the set of dykes that are not broken. Therefore, the total number of edges is given by

$$\begin{aligned} E &= \text{(the number of unbroken dykes)} + \text{(the number of broken dykes)} \\ &= (V - 1) + (F - 1) = V + F - 2. \end{aligned}$$

Therefore, $\quad V - E + F = 2.$

Exercise.

(a) Show that there exists no convex solid with seven edges.

(b) However, for any integer $k \geq 6$, but $k \neq 7$, there exists a convex solid with k edges.

(c) Consider the solid obtained by removing the unit cube at the center of the $3 \times 3 \times 3$ cube. Is the Euler formula true for this solid? If not, where did our proof go wrong? And what condition must be imposed on a solid in order for the Euler formula to be valid? What if we remove two more unit cubes (i.e., all together we remove three unit cubes) each sharing a face with the one at the center?

15.2 Regular Polyhedra

As an application of the Euler formula, we now classify all the regular polyhedra. But which polyhedron qualifies to be called *regular*? It appears to be reasonable to require:
(a) all the faces are congruent regular polygons (of n sides, say), and
(b) at all the vertices, a same number (say, p) of edges meet.
It turns out that these two requirements are sufficient to imply that no more than 5 types of regular polyhedra exist.

Because at each of the V vertices, p number of edges meet, we have

$$2E = pV, \quad V = \frac{2E}{p}.$$

(Note that each edges is counted twice, once at each end.) Similarly, two faces meet at each of the edges, and each face has n edges, we obtain

$$nF = 2E, \quad F = \frac{2E}{n}.$$

Substituting these two equalities into the Euler formula $V - E + V = 2$, we obtain

$$\frac{2E}{p} - E + \frac{2E}{n} = 2.$$

Dividing both sides by $2E$, we obtain

$$\frac{1}{p} + \frac{1}{n} = \frac{1}{2} + \frac{1}{E}.$$

Because n is the number of the sides of a polygon, we must have $n \geq 3$. Similarly, p is the number of the edges that meet at a vertex, and so $p \geq 3$. However, at least one of n and p must be 3, otherwise, $n \geq 4$ and $p \geq 4$ would imply

$$\frac{1}{p} + \frac{1}{n} \leq \frac{1}{4} + \frac{1}{4} = \frac{1}{2} < \frac{1}{2} + \frac{1}{E},$$

The Euler Formula

as E must be positive. Now suppose $n = 3$, then

$$\frac{1}{p} = \frac{1}{2} - \frac{1}{n} + \frac{1}{E} = \frac{1}{2} - \frac{1}{3} + \frac{1}{E} = \frac{1}{6} + \frac{1}{E} > \frac{1}{6},$$

implying that $p < 6$. Because the equality $\frac{1}{p} + \frac{1}{n} = \frac{1}{2} + \frac{1}{E}$ is symmetric with respect to p and n, we conclude that at least one of n and p must be 3, and

$$3 \le p \le 5, \quad 3 \le n \le 5.$$

For $n = 3$, we obtain

$$\frac{1}{p} = \frac{1}{6} + \frac{1}{E}.$$

Substituting $p = 3, 4, 5$, we obtain $E = 6, 12, 30$, respectively, corresponding to the tetrahedron, octahedron, and icosahedron.

Similarly, for $p = 3$, we obtain

$$\frac{1}{n} = \frac{1}{6} + \frac{1}{E}.$$

Substituting $n = 3, 4, 5$, we obtain $E = 6, 12, 30$, respectively, corresponding to the tetrahedron, cube, and dodecahedron. Finally, substituting the values of n, p, and E into

$$V = \frac{2E}{p}, \quad F = \frac{2E}{n},$$

we obtain the numbers of the vertices and the faces in the corresponding polyhedra.

n	p	V	E	F	polyhedron
3	3	4	6	4	tetrahedron
4	3	8	12	6	cube
3	4	6	12	8	octahedron
5	3	20	30	12	dodecahedron
3	5	12	30	20	icosahedron

Note that comparing the cube and the octahedron, we see that the values of n and p are interchanged, and so are the values of V and F. Similarly, for the dodecahedron and the icosahedron.

Exercise. A soccer ball is covered by "pentagons" and "hexagons". How many are there for each type? Note that to each pentagon, there are 5 hexagons adjacent to it, while to each hexagon, there are three pentagon adjacent to it.

Chapter 16
The Sperner Lemma

16.1 The Sperner Lemma

In this chapter, we present a result that is not only a gem in itself, but it has wonderful proofs. Furthermore, it has a nice application to the proof of the important Brouwer fixed point theorem.

Choose any finite number of points in the interior and boundary of a given triangle ABC. Join those pairs of points that can be joined without crossing segments that are already drawn (except, of course, at the endpoints), nor going through one of the chosen points. If this construction is carried out as far as possible, we obtain a partition of the original triangle into "subtriangles" with the property that any pair of subtriangles have (i) no common point, (ii) one common vertex, or (iii) one common side. Such a partition into subtriangles is called a *triangulation* of the original triangle ABC. (Figure SPL/Left) Note that Figure SPL/Right is not allowed. (It violates condition (iii)).

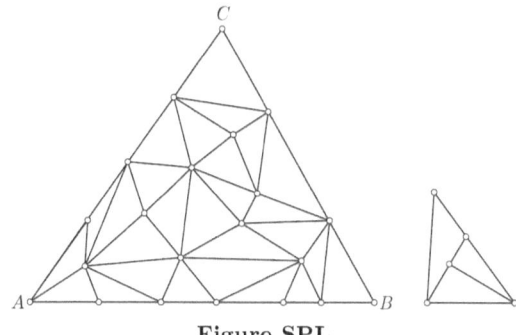

Figure SPL

Exercise. Show that the number of subtriangles in a triangulation of a triangle is completely determined by the numbers of points (vertices) in the interior and on the boundary only. What if the triangle is replaced by a convex quadrangle? A convex pentagon, etc.? What if the convexity in the assumption is deleted?

Lemma 1. [Sperner] Given a triangulation of $\triangle ABC$, we label each vertex of the subtriangles A, B, or C at random subject to the boundary condition that a vertex on a side of the original triangle ABC must be labelled by one of the labels at the endpoints of this side. Then the number of subtriangles that is labelled ABC is odd. In particular, there exists at least one such subtriangle.

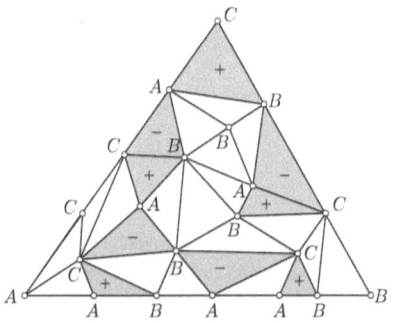

Figure SPL1

We prove a stronger version. Divide the set of all those triangles that are labelled ABC into two subsets depending on whether they are labelled A-B-C clockwise or counterclockwise. Then we claim: the subset, the orientation of whose elements agrees with that of the original triangle, contains exactly one more element than the other subset. (Hence together there are odd number of them.) In Figure SPL1, the triangles labelled counterclockwise and clockwise ones are indicated by "+" and "−", respectively.

Proof. The proof is surprisingly simple. To fix our idea, we assume that the original triangle ABC is labelled A-B-C counterclockwise.

First, to each side of a subtriangle in the triangulation, we assign a "weight":

(a) The weight of a side is 0 if two ends have the same label.

(b) The weight of a side is 1 if, on moving along the side of the triangle counterclockwise, the side is labelled BC, CA, or AB.

(c) The weight of a side is −1 if, on moving along the side of the triangle counterclockwise, the side is labelled CB, AC, or BA.

The weight of a subtriangle in the triangulation is the sum of the weights of its three sides. (Figure SPL2)

The Sperner Lemma 113

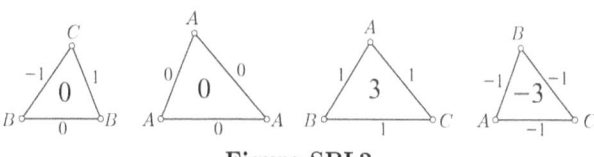

Figure SPL2

Then the weight of a subtriangle is 3 if it is labelled A-B-C counterclockwise, and -3 if clockwise. All other subtriangles have the weight 0. Hence our proof is complete if we show that the total sum of all the weights of subtriangle is 3.

Clearly, the contribution to the total sum by any segment that is in the interior of the original triangle is 0 because such a segment is traversed twice, once in each direction. However, if a segment is on the side of the original triangle, then it is traversed only once. Thus we reduced our proof of the 2-dimensional version of the Sperner lemma to that of the 1-dimensional version.

Lemma 2. Suppose that a finite number of points divide a closed interval $[A, B]$ into subintervals, and each of the partition points is also labelled A or B. Then the number of subintervals whose endpoints are labelled by different letters (i.e., subintervals labelled $[A, B]$ or $[B, A]$) is odd. In particular, there exists at least one such subinterval.

To each closed subinterval we assign a weight by first assign a weight for its endpoints as follows (Figure SPL3): point A carries weight 0 (whether it appears as the left endpoint or the right endpoint of the subinterval); point B carries weight 1 if it appears as the right endpoint, but -1 if it appears as the left endpoint of the subinterval. And the weight of a subinterval is the sum of the weights at its two endpoints. Thus

subinterval $[A, A]$ has weight 0; subinterval $[A, B]$ has weight 1;
subinterval $[B, A]$ has weight -1; subinterval $[B, B]$ has weight 0.

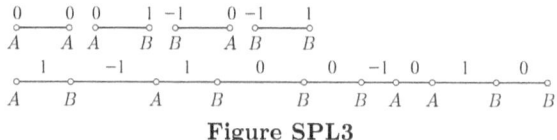

Figure SPL3

Now consider the sum of all the weights of the subintervals. By definition, it is equal to the sum of all the weights at all the points. But the sum of the two weights at the two sides of any interior point (whether it is labelled A or B) is 0, while the weights at the two extreme endpoints are 0 and 1, respectively. Thus the total sum is 1, implying that the number of the subintervals labelled $[A, B]$ is one more than that of the subintervals labelled $[B, A]$. This completes our proof of the 1-dimensional version.

It follows that the sum of all the weights on side $[A, B]$ of the original triangle is 1. Similarly, for the other two sides, BC and CA, of the original triangle; and our proof (for the 2-dimensional version) is also complete. □

We now propose the following game. (Actually, I find this is a great way to start a talk to get students involved and arose their interest. Furthermore, I challenge the two volunteer students whether they can cooperate so that the game ends in a tie.)

1. Two players take turns to place points in the interior and boundary of a given triangle ABC until the number of points reaches the pre-agreed number.

2. Next, the players take turns to draw segments until a triangulation of $\triangle ABC$ is achieved (with the vertices chosen in step 1).

3. Then the players take turns to label each of the vertices (of the subtriangles) A, B, or C subject to the boundary condition mentioned above.

4. The player who *completed* the labelling of any one of the subtriangles as ABC is the loser (regardless of which player labelled the first two vertices of that subtriangle).

5. A variation of this game is to keep on playing until all the vertices are labelled. The player who completed labelling of more subtriangles as ABC is the loser.

The Sperner lemma guarantees that either version of this game never ends in a tie. (I would appreciate it very much if a reader can show me a winning strategy.)

Proof. Our second proof of the Sperner lemma can be described as a "walking process". We consider

the original triangle ABC as a "house",
each of the subtriangles as a "room", and
the space outside the house as a "yard".

And we install a "door" in each of the segments whose two endpoints are labelled A and B. Then we have three types of rooms (Figure SPL4):

Type	Example
0-door room	(AAA), (CCC), (ACC), (BBC),
1-door room	(ABC)
2-door room	(AAB), (ABB)

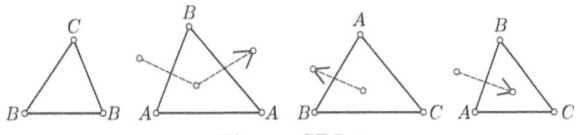

Figure SPL4

Now each walk must satisfy the following conditions:

The Sperner Lemma

(a) Any door can be passed at most once;

(b) Every walk must starts from a 1-door room and continue until there is no room or yard to walk into; in particular, when the path leads to the yard, and if there is a door that has not been passed through on side AB of the original triangle, then we must continue our walk and go into the house.

Consequently, upon entering a 2-door room, we must exit from the other door. Moreover, once the starting 1-door room is chosen, the walk path is uniquely determined. After the walk comes to the end, we start another one, and repeat this process until there is no more 1-door room left to start from. (Figure SPL5) Note that there can

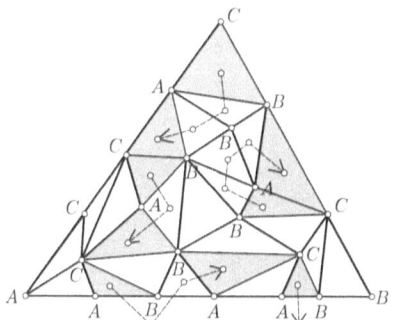

Figure SPL5

be at most one path that ends in the yard, because any two paths that end in the yard should be combined into a single path according to condition (b). Clearly, a path ends either in a 1-door room or in the yard. Each path that ends in a 1-door room corresponds to a pair of 1-door rooms (and also even number of doors on side AB of the original triangle). But a path that ends in the yard corresponds to a single 1-door room (and an odd number of doors on side AB of the original triangle).

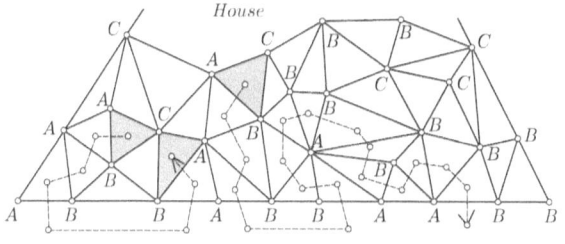

Figure SPL6

Therefore, our proof is complete if we show that the number of doors on side AB

of the original triangle is odd. (Equivalently, there must be a path that ends in the yard). So again we reduced our proof to the 1-dimensional case.

In the 1-dimensional case, our "house" is the original interval $[A, B]$; a "room" is a subinterval; and the "yard' is the half-line $[B, \infty)$ outside the house. And we install a "door" at each point labelled B (but not at A's). Thus subinterval $[A, A]$ is a 0-door room; $[A, B]$ and $[B, A]$ are 1-door rooms; and $[B, B]$ is a 2-door room. (Figure SPL7) As before, our walk must satisfy the following conditions:

(a) Any door can be passed at most once;

(b) Every walk must starts from a 1-door room and continue until there is no room or yard to walk into.

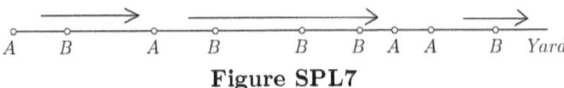

Figure SPL7

Consequently, upon entering a 2-door room, we must exit from the other door; and once the starting 1-door room is chosen, then the walk path is uniquely determined. After the walk is finished; i.e., no more room or yard to walk into, then we start another one. We repeat this process until there is no 1-door room left to start from. Clearly, there exists one and only one path that ends in the yard because there is one and only one door that leads to the yard at the extreme right endpoint B, and by considering walking backward from the yard, we see there must be a 1-door room from which a walk path leads to the yard. Obviously, our walk ends either in a 1-door room or in the yard. Each path that ends in a 1-door room corresponds to a pair of 1-door rooms, while the path that ends in the yard corresponds to a single 1-door room. Hence the number of 1-door room must be odd, and this completes our proof of the 1-dimensional case as well as that of the 2-dimensional case. (Note that the existence of at least one walk in the 2-dimensional case is guaranteed by the 1-dimensional version.) □

Alternatively, for the 1-dimensional case we can reason as follows: Let one side of the line AB be called a "house" and the other side the "yard." (Figure SPL8) We install a one-way door from the house to the yard in each interval labelled $[A, B]$, and a one-way door from the yard to the house in each interval labelled $[B, A]$, but no door of any kind in intervals labelled $[A, A]$ or $[B, B]$. (They are "walls".)

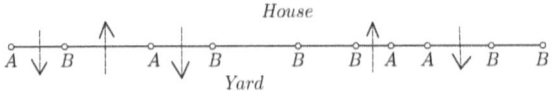

Figure SPL8

Clearly, we cannot have two consecutive doors that are going the same direction. Hence the doors alternate "in" and "out". And the first and the last doors (these two may coincide) must be going the same direction because the two extreme endpoints are labelled by different letters, A and B. As the directions of the doors alternate, the number of doors must be odd, and that completes our proof of the 1-dimensional case. Note that we solved a 1-dimensional problem by considering it in a 2-dimensional space.

Meticulous readers might think that the second proof is inferior compared to the first one because it does not prove the numbers of subtriangles labelled A-B-C counterclockwise and clockwise differ by 1. But this defect can easily be remedied. We give a hint, but leave the details to the reader; i.e., "I do the easy part, and you do the hard part." All we need is to change all the doors to one-way doors; i.e., modify the doors so that when going through the door, we see A on the right, and B on the left. And we require the starting room to be labelled A-B-C counterclockwise. Then the path must end either in a room labelled A-B-C clockwise (because throughout the walk, we see nothing but A's on the right and B's on the left, so the ending room must be labelled A-B-C clockwise) or the yard. Clearly, after all the walks, it is impossible to have only rooms labelled A-B-C clockwise are left with no room labelled A-B-C counterclockwise to start a new walk. (Why?)

Exercise. What about the 3-dimensional version of the Sperner lemma?

16.2 The Brouwer Fixed Point Theorem

As an application of the Sperner lemma, we present the Brouwer fixed point theorem for a triangle (which is topologically equivalent to a disc). Note that our "triangle" is the union of its interior and its boundary.

> *Theorem.* [Brouwer] A continuous function f that maps a triangle into itself has a fixed point; i.e., there exists a point P in the triangle such that $f(P) = P$.

Proof. Through a point O in the plane of $\triangle ABC$, draw lines parallel to the three sides BC, CA, AB of $\triangle ABC$ dividing the plane into 6 angular regions. We call three of them as region A, region B, region C, respectively. (Shaded in Figure SPL9/Right) And in the three opposite regions, draw half-lines OX, OY, OZ, subject to the condition that each of $\angle YOZ$, $\angle ZOX$, $\angle XOY$ is less than π. (Clearly, this is possible for any $\triangle ABC$. Simply extend the angle bisectors of the regions A, B, C to the respective opposite angular regions, for example.)

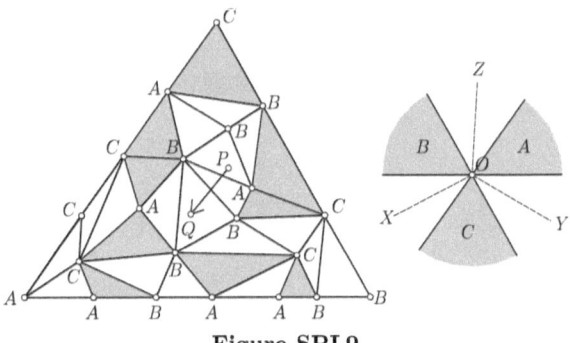

Figure SPL9

Suppose there does not exist a fixed point. Then for any point P in $\triangle ABC$, $\overrightarrow{PQ}$, where $Q = f(P)$, is not a zero vector, and so vector $\overrightarrow{PQ}$ has a direction. Hence we may call point P an A-point, B-point, or C-point, depending on whether the direction of $\overrightarrow{PQ}$ is inside the angular region $\angle YOZ$, $\angle ZOX$, or $\angle XOY$. If this direction coincides with one of the (boundary) directions, say $\overrightarrow{OX}$, we may call it as a B-point or C-point at random. Or, if we wish, we could declare all such points to be B-points. And similarly, all the points that have the same directions as $\overrightarrow{OY}$ or $\overrightarrow{OZ}$ as C-points or A-points, respectively.

Note that vertices A, B, C of $\triangle ABC$ are an A-point, B-point, C-point, respectively, while a point on side AB (but not one of the two endpoints) must be either an A-point or a B-point. Similarly, for the points on sides BC and CA. And this classifies every point in the (closed) triangle into one of the three categories (under our assumption that no fixed point exists).

We now consider a sequence $\{T_n\}_{n=1}^{\infty}$ of triangulations of $\triangle ABC$ such that $\delta(T_n) \longrightarrow 0$ (as $n \to \infty$), where $\delta(T_n)$ is the maximal length of all the sides of all the subtriangles in triangulation T_n. (Clearly, such sequence of triangulations exists.) By the Sperner lemma, for each n, there exists a subtriangle $A_n B_n C_n$ such that A_n is an A-point, B_n is a B-point, and C_n is a C-point.

Now sequence $\{A_n\}_{n=1}^{\infty}$ of points may not converge, but because $\triangle ABC$ is closed and bounded, there must be a convergent subsequence. So, after replacing the sequence $\{T_n\}_{n=1}^{\infty}$ by its suitable subsequence (which, for simplicity, we also denote by $\{T_n\}_{n=1}^{\infty}$; and correspondingly, for $\{A_n\}_{n=1}^{\infty}$, $\{B_n\}_{n=1}^{\infty}$, and $\{C_n\}_{n=1}^{\infty}$), we may assume that $\{A_n\}_{n=1}^{\infty}$ converges, say to point P_0, in $\triangle ABC$. Because the distances of B_n and C_n from A_n are at most $\delta(T_n)$, which converges to 0, the sequences $\{B_n\}_{n=1}^{\infty}$ and $\{C_n\}_{n=1}^{\infty}$ also converge to the same point P_0.

Because we are assuming that our function f does not have a fixed point, $\overrightarrow{P_0 Q_0}$ (where $Q_0 = f(P_0)$) has a direction. Now consider (the absolute values of) the three angles that $\overrightarrow{P_0 Q_0}$ makes with $\overrightarrow{OX}$, $\overrightarrow{OY}$, $\overrightarrow{OZ}$. Suppose, of these three angles, the angle between $\overrightarrow{P_0 Q_0}$ and $\overrightarrow{OZ}$ is a minimum. (Other cases are similar, and our argument below works even if two of these angles are equal and minimum.)

The Sperner Lemma

Now the continuity of f on $\triangle ABC$, in particular, at point P_0 means: given a neighborhood W of $Q_0 = f(P_0)$, no matter how small, there exists a neighborhood V of P_0 such that all the images of points in V must be in W. Because the angle between $\overrightarrow{P_0 Q_0}$ and $\overrightarrow{OZ}$ is a minimum, we can choose W and V (first W, then V) sufficiently small such that none of the point in V is an C-point. (Loosely speaking, the direction of any vector with its initial point in V and the terminal point in W cannot be too far off from that of $\overrightarrow{P_0 Q_0}$ when both V and W are sufficiently small by the continuity of f.) (Figure SPL10)

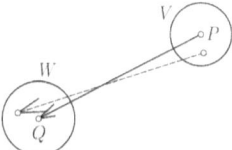

Figure SPL10

But sequence $\{C_n\}_{n=1}^\infty$ is a sequence of C-points that converges to P_0, and V is a neighborhood of P_0, so V must contain all the points in some tail part of sequence $\{C_n\}_{n=1}^\infty$. This contradicts our assertion above. We obtain a contradiction because we assume that f does not have a fixed point; our proof is now complete. □

Exercise. Suppose $\triangle XYZ \sim \triangle ABC$ (X corresponds to A, Y to B, Z to C), and $\triangle XYZ$ is completely inside $\triangle ABC$. Locate the fixed point (i.e., find point P whose relative positions to $\triangle XYZ$ and to $\triangle ABC$ are the same).

16.3 An Elementary Fixed Point Theorem

In this section, we present an example of the fixed point theorem for a similarity transformation so that we can find the fixed point by a compass-straightedge construction. Here is the problem we want to solve.

> Suppose two rectangular maps $ABCD$ and $A'B'C'D'$, identical except in size, are placed one on top of the other so that the small one does not stick out from the big one. How do we find a point that has the same latitude and longitude in the maps? (Figure FXP)

Naturally, a similar problem can be posed if maps are replaced by photos.

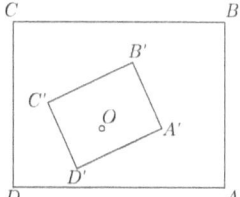

Figure FXP

Note that a pair of corresponding segments, say, diagonals BD and $B'D'$, even if they intersect each other inside the small rectangle, their intersection may not divide two segments into the same ratio, in which case the desired fixed point is not on either one of these segments. On the other hand, if a segment in one of the rectangles passes through the fixed point, then its corresponding segment in the other rectangle must also pass through the fixed point, and their intersection, being the fixed point of a similarity transformation, must divide the two segments into the same ratio. Converse is trivial. If a pair of corresponding segments that intersect each other and their intersection divides the segments into the same ratio, then the intersection must be the fixed point.

The following (which is discussed in every elementary geometry textbook) is all we need. We leave the proof as an exercise for the reader.

Theorem. Let A', B' be points on sides OA, OB (or their extensions) of $\triangle OAB$, respectively. (Figure FXP1) Then

$$A'B' \parallel AB \implies \frac{\overline{OA'}}{\overline{OA}} = \frac{\overline{A'B'}}{\overline{AB}} = \frac{\overline{OB'}}{\overline{OB}}.$$

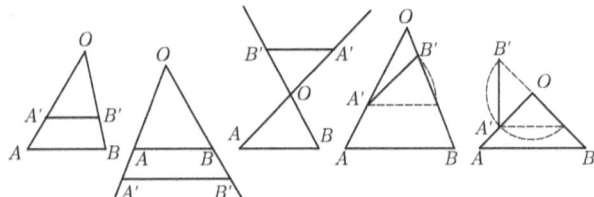

Figure FXP1

Exercise. Is the converse true?

Figure FXP2 is self-explanatory. It shows how to construct a pair of points P and P' on two given segments AB and $A'B'$, respectively, such that

$$\frac{\overline{AP}}{\overline{PB}} = \frac{\overline{A'P'}}{\overline{P'B'}} \quad \text{and} \quad PP' \perp AB.$$

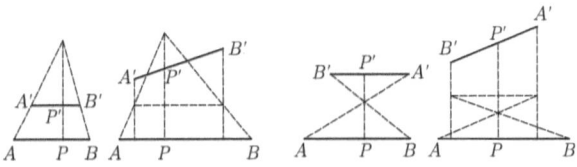

Figure FXP2

The Sperner Lemma

So, in the given configuration (Figure FXP3), we find points P, P', Q, Q' on AB, $A'B'$, CD, $C'D'$, respectively, such that

$$\frac{\overline{AP}}{\overline{PB}} = \frac{\overline{A'P'}}{\overline{P'B'}}, \quad PP' \perp AB; \quad \frac{\overline{CQ}}{\overline{QD}} = \frac{\overline{C'Q'}}{\overline{Q'D'}}, \quad QQ' \perp CD.$$

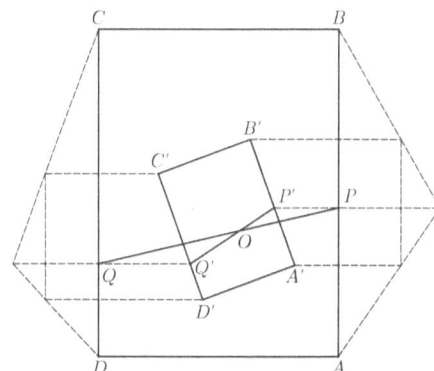

Figure FXP3

Then obviously, PQ and $P'Q'$ have the same relative position in each of the rectangles. Furthermore, $PP' \perp AB \parallel CD \perp QQ'$, and so $PP' \parallel QQ'$; and because rectangle $A'B'C'D'$ is inside rectangle $ABCD$, vectors $\overrightarrow{PP'}$ and $\overrightarrow{QQ'}$ have opposite direction, hence segments PQ and $P'Q'$ do intersect, say at O. Then clearly, we have $\triangle OPP' \sim \triangle OQQ'$ (because $PP' \parallel QQ'$, and so $\angle OPP' = \angle OQQ'$, and $\angle POP' = \angle QOQ'$), implying

$$\frac{\overline{PO}}{\overline{OQ}} = \frac{\overline{P'O}}{\overline{OQ'}},$$

i.e., the intersection O divides the pair of corresponding segments PQ and $P'Q'$ into the same ratio. So our solution is complete.

Exercise.

(a) What if PQ and $P'Q'$ are on a same straight line?

(b) What if $A'B' \perp AB$?

(c) What if two maps are facing each other? (Figure FXP4)

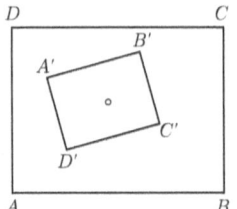

Figure FXP4

So far so good. But if the rectangles are replaced by other shapes such as triangles or discs? Actually, a pair of points and their images completely determine a similarity transformation. Hence all we need is to solve the following problem:

Given four points A, B, A', B' in a plane, find the point O such that
$$\triangle OAB \sim \triangle OA'B'.$$

(Note that we require A and B to correspond to A' and B', respectively. Moreover, two triangles must have the same orientation; i.e., labels O-A-B and O-A'-B' are either both counterclockwise or both clockwise.) (Figure FXP5)

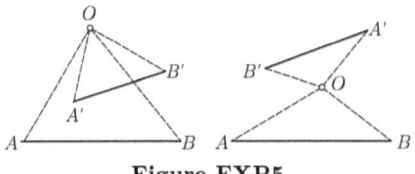

Figure FXP5

Let T be the point such that quadrangle $BB'A'T$ is a parallelogram. (Figure FXP6) We claim:

Point O that satisfies $\triangle ATB \sim \triangle AA'O$ is the desired point.

Because $\triangle AA'O$ is obtained by a stretching-contraction of $\triangle ATB$ followed by a rotation around point A, and so the angles between the corresponding sides are all equal. In particular, the angle between the sides AO and AB (i.e., $\angle OAB$) is equal to the angle between the sides $A'O$ and TB. But $TB \parallel A'B'$, hence $\angle OAB = \angle OA'B'$. Furthermore, because $\triangle ATB \sim \triangle AA'O$, we have

$$\frac{\overline{AO}}{\overline{AB}} = \frac{\overline{A'O}}{\overline{TB}} = \frac{\overline{A'O}}{\overline{A'B'}}.$$

The Sperner Lemma

Therefore, $\triangle OAB \sim \triangle OA'B'$ (SAS), which justifies our construction.

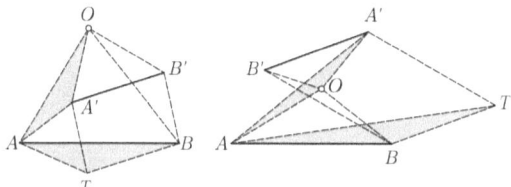

Figure FXP6

Exercise. Given four points A, B, A', B' in a plane, find the point O such that

$$\triangle OAB \sim \triangle OA'B' \quad \text{(reversed)}.$$

The following solution is essentially due to Mr. Dean Ballard of Lakeside School, Seattle.

Suppose a pair of points A, B, is mapped to a pair of points A', B', by a similarity transformation; and point P is the fixed point. (Figure FXP7) Then

$$\triangle PA'B' \sim \triangle PAB,$$

which implies

$$\triangle PA'A \sim \triangle PB'B \quad (SAS).$$

In particular,

$$\angle PA'A = \angle PB'B;$$

equivalently,

$$\angle PA'Q = \angle PB'Q,$$

where Q is the intersection of lines AA' and BB'. Therefore, points P, Q, A', B' are cocyclic.

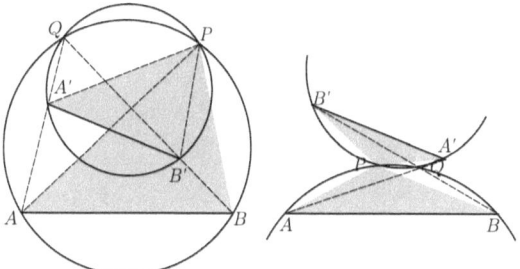

Figure FXP7

Similarly,
$$\triangle PA'A \sim \triangle PB'B$$
also implies that points P, Q, A, B are cocyclic. We have shown that P is the (second) intersection of the circumcircles of $\triangle QA'B'$ and $\triangle QAB$.

Chapter 17

Lattice Points

17.1 The Pick Theorem

Lattice points are points whose coordinates are integers. A *simple lattice polygon* is a polygon that is not self-intersecting and all of whose vertices are lattice points. A beautiful theorem of Georg Alexander Pick asserts that the area of a simple lattice polygon is completely determined by the numbers of the lattice points in the interior and on the boundary of the polygon. (Figure PCK)

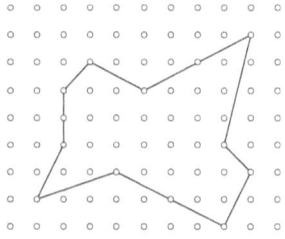

Figure PCK

Now suppose you heard about the existence of such a theorem, how do you go about finding the area formula of a simple lattice polygon in terms of the numbers of its lattice points in the interior and on its boundary? Naturally, we start by experimenting with easy cases. Take an isosceles right triangle which is half of a unit square. (Figure PCK1) Obviously, its area is $\frac{1}{2}$. If we add one more boundary (lattice) point (without introducing an interior lattice point), then regardless of how we do it, the area increases by $\frac{1}{2}$. So it is only natural that the coefficient of b (the number of the lattice points on the boundary) must be $\frac{1}{2}$.

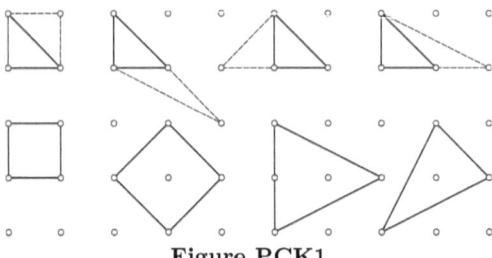

Figure PCK1

But what about the coefficient of i (the number of lattice points in the interior)? Compare the two smallest lattice squares; i.e., the unit square and the square with side length $\sqrt{2}$. (Figure PCK1) Either one of them has four boundary points, but the latter has one (extra) interior lattice point, and that resulted in the increase of the area from 1 to 2. So we figure that the coefficient of i should be 1.

Hence our first guess of the formula is $i + \frac{1}{2}b$. (By sticking two lattice polygons together, we see it is unlikely that there will be a term such as ib in the area formula.) For the isosceles right triangle which is half of the unit square, we have $i = 0$, $b = 3$, and so the formula gives $\frac{3}{2}$, but of course the actual area is $\frac{1}{2}$. For the unit square, we have $i = 0$, $b = 4$, and so the formula gives 2, but the actual area is 1. For the square with side length $\sqrt{2}$, we have $i = 1$, $b = 4$, and so the formula gives 3, but the actual area is 2.

In all these cases, our formula over estimates the area by 1. So we add the error adjusting term -1, to obtain the following.

Conjecture. The area of an arbitrary simple lattice polygon is

$$i + \frac{b}{2} - 1,$$

where i and b are, respectively, the numbers of the lattice points in the interior and on the boundary of the polygon.

Let us test our formula for an $m \times n$ lattice rectangle (where m and n are positive integers). Clearly, the number i of the interior lattice points is $(m-1)(n-1)$; and the number b of the boundary lattice points is $2(m+n)$. (Figure PCK2) So our formula gives the area as

$$(m-1)(n-1) + \frac{1}{2} \cdot 2(m+n) - 1 = mn,$$

which is correct. Now we have more confidence in our formula.

Lattice Points

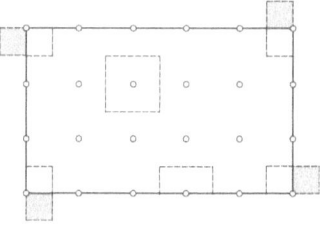

Figure PCK2

In fact, by attaching a unit square to each lattice point (Figure PCK2), we are confident that our formula is the only sensible choice if there ever is such a formula. Even for other polygons (Figure PCK3), we see immediately that the coefficient of i should be 1 (as long as the attached unit square is entirely inside the lattice polygon). Similarly, for b, the coefficient should be 1/2 (as long as the boundary lattice point is not one of the vertices of the polygon), because a line passing through the center of a square always bisects the area of the square.

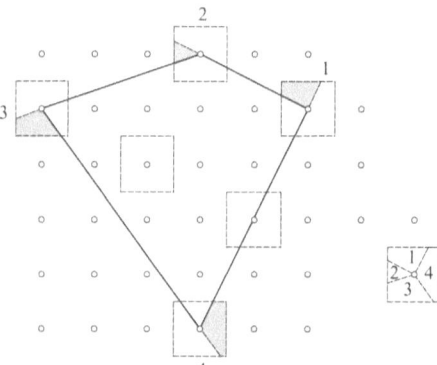

Figure PCK3

As for vertices, it is not difficult to see that, with coefficient 1/2, we "over counted" if the interior angle at the vertex is less than 180° (i.e., if the exterior angle is positive), but "under counted" if the interior angle there is more than 180° (i.e., if the exterior angle there is negative). (Figure PCK4) However, the algebraic sum of the exterior angles of a (simple) polygon is 360° and those over counted parts can be assembled (in fact, parallel translations will do) to form a unit square. Hence we must subtract the area of a unit square. This justifies the constant term -1 in the formula again.

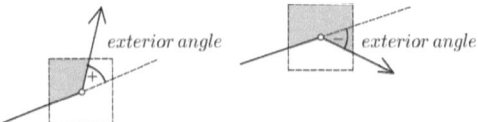

Figure PCK4

But how do we justify for those interior lattice points whose attached unit squares are not entirely inside the polygon? It is easy to see that if the attached unit square U of an interior lattice point X is cut by side AB (where A and B are lattice points) of the polygon into two pieces, then there exists a lattice point X' that is symmetric to X with respect to midpoint M of segment AB (hence on the opposite side of AB) whose attached unit square U' (symmetric to U with respect to M) is also cut by AB into two pieces exactly the same way as U. (Figure PCK5) Therefore, the "missing" part of U is compensated precisely by the "extra" part from U'.

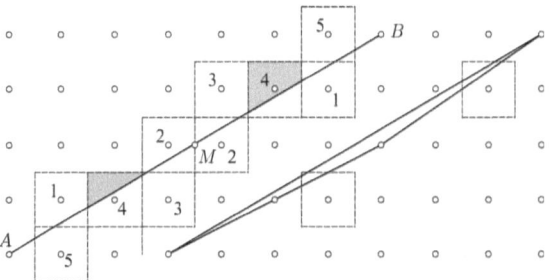

Figure PCK5

All seems to be going well, but what if an attached unit square is cut twice (or more) by the boundary of the polygon such as in the case of a very skinny triangle? (Figure PCK5) Perhaps with some effort, we can overcome this difficulty, but the solution is unlikely to be very clean. So let us try a fresh approach.

Let us call a triangle *primitive* if its vertices are lattice points, but it does not have any other lattice points either in its interior or on its sides. Any lattice polygon P can be "triangulated" into primitive triangles (i.e., can be subdivided into a finite number of primitive subtriangles) — in fact, in many ways. (Figure PCK6)

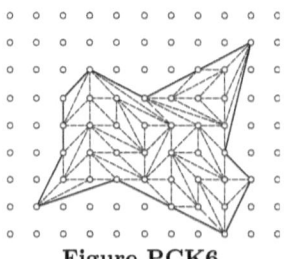

Figure PCK6

Lattice Points

Our conjecture will be justified if we can prove the following.

Lemma. (a) The area of any primitive triangle is $\frac{1}{2}$.
(b) In any partition of a simple lattice polygon P into primitive subtriangles, the number n of the primitive triangles is
$$n = 2i + b - 2,$$
where i and b are the numbers of lattice points in the interior and on the boundary, respectively, of polygon P. In other words, the number n of the primitive subtriangles in a partition depends only on the numbers i and b.

We start with the proof of part (b) by calculating the sum of all the angles of the primitive triangles in the partition.

Each interior lattice point of P is a common vertex of some primitive subtriangles in the partition. And the angles at each of such interior lattice points sum up to 360°. Because polygon P has i interior lattice points, their total contribution is $i \times 360°$.

Let v be the number of vertices of polygon P. At each lattice point on the boundary that is not a vertex of polygon P, the angles of primitive triangles sum up to 180°. And because P has $(b-v)$ such lattice points, their total contribution is $(b - v) \times 180°$.

Finally, the sum of the interior angles of any simple polygon, convex or not, with v vertices is $(v-2) \times 180°$.

So, if the partition of polygon P consists of n primitive subtriangles, then
$$n \times 180° = i \times 360° + (b-v) \times 180° + (v-2) \times 180°\,; \text{ i.e., } n = 2i + b - 2.$$

This completes the proof of part (b).

Note carefully, in our proof here, the assumption that those interior points and boundary points are lattice points is not used. Hence we have the following.

The number of subtriangles in a triangulation of a simple polygon is
$$n = 2i + b - 2,$$
where i is the number of chosen points in the interior, and b is that of the chosen points on the boundary (including the vertices) of the polygon.

We might as well give an alternate proof. Recall that a triangulation of a polygon is a partition of the polygon into subtriangles (see §16.1 *The Sperner Lemma*, page 111), such that any pair of subtriangles have (i) no common point, (ii) one common vertex, or (iii) one common side.

Observe that given a partition of a polygon, if we add one more interior point, then this new point must be either in the interior of one of the subtriangles (as point A in Figure PCK7), or on the side of one of the subtriangles (as point B in Figure PCK7). In either case, it is easy to see that the addition of a new interior point causes the number of subtriangles to increase by 2 (because in the case of point A, one subtriangle is replaced by 3; while for point B, two subtriangles become four). Hence the coefficient of i must be 2.

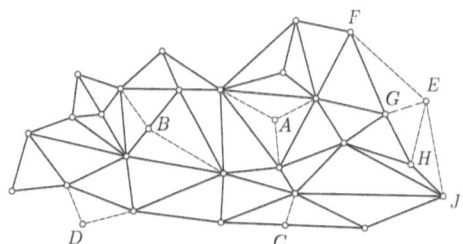

Figure PCK7

Similarly, there are two ways to add a new boundary point. One way is that the new boundary point is on the boundary of the polygon (as point C in Figure PCK7). Another way is that this new boundary point is outside the polygon (as points D and E in Figure PCK7). Each of the cases of C and D clearly increases the number of subtriangles by 1. Hence, clearly, it is only natural to set the coefficient of b as 1. As for the case of E, we may consider joining E with points F and G first, that results in the increase of the number of subtriangles by 1 (as in the case of point D). Then join E and H, that increases the number of subtriangles by 1, and turns the boundary point G into an interior point, which justifies the change of the corresponding coefficient from 1 to 2. Now simply repeat the same reasoning as we join E and J, etc.

Clearly, any configuration of a triangulation of a polygon can be constructed by adding boundary points one by one starting from any one of the subtriangles. And for any triangle, we have $i = 0$, $b = 3$; so with coefficients 2 and 1 for i and b, respectively, we need an error adjusting term -2. Therefore, the correct formula is

$$n = 2i + b - 2.$$

Exercise. Prove the Euler formula $V - E + F = 2$ (page 107) by imitating the reasoning above.

It remains to prove part (a). Given any lattice triangle, primitive or not, we may attach lattice right triangles (whose legs are parallel to two coordinate axes) such that their union is a lattice rectangle (whose sides are parallel to the coordinate axes). (Figure PCK8) Clearly, the area of any lattice rectangle is an integer, and that of each of the lattice right triangles is either an integer or half of an integer. Therefore, the area of the given lattice triangle must be either an integer or half of an integer. In particular,

$$\text{the area of any lattice triangle is at least } \frac{1}{2}.$$

Lattice Points

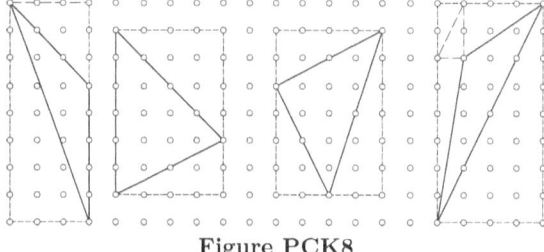

Figure PCK8

Readers who have studied determinants may notice that this assertion also follows from the formula that the (signed) area of a triangle whose vertices are at (x_1, y_1), (x_2, y_2), (x_3, y_3) is

$$\frac{1}{2} \begin{vmatrix} x_1 & y_1 & 1 \\ x_2 & y_2 & 1 \\ x_3 & y_3 & 1 \end{vmatrix}.$$

Given an arbitrary primitive triangle T, consider an arbitrary lattice rectangle Q (whose sides are parallel to the coordinate axes) that contains T. (Figure PCK9) Then partition rectangle Q into primitive subtriangles using T as one of the primitive subtriangles. Such a partition exists because T itself is primitive. Suppose the size of rectangle Q is $m \times n$. Then clearly, Q has $(m-1)(n-1)$ interior lattice points and $2(m+n)$ boundary lattice points. Therefore, by part (b) of the lemma (which we have already established), the number of the primitive subtriangles is

$$2(m-1)(n-1) + 2(m+n) - 2 = 2mn.$$

Now the area of any lattice triangle is at least $\frac{1}{2}$, hence the area of Q is at least mn.

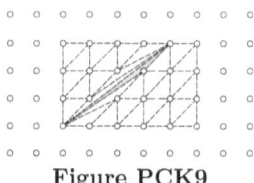

Figure PCK9

But clearly, the area of Q is precisely mn. It follows that the area of every primitive subtriangle (in our partition), including T, must be $\frac{1}{2}$.
This completes the proof of our lemma and we obtain the following.

Theorem. [G. Pick] The area of an arbitrary simple lattice polygon is

$$i + \frac{b}{2} - 1,$$

where i and b are, respectively, the numbers of the lattice points in the interior and on the boundary of the polygon.

Exercises.

(a) How do you explain that by simply bisecting a 13 × 5 rectangle or a 13 × 11 rectangle along a diagonal into a pair of right triangles (Figure PCK10), one of the right triangles gets an extra (unit) square?

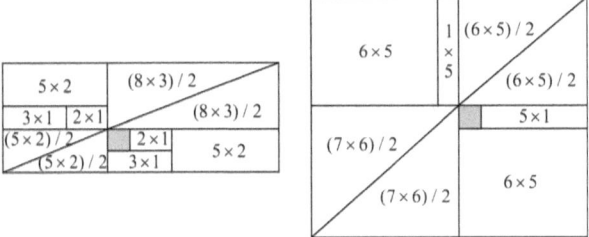

Figure PCK10

(b) Show that given a triangle whose three sides have integer lengths and the area is also an integer, we can always find a lattice triangle congruent to the given triangle. What if the area is one half of an (odd) integer?

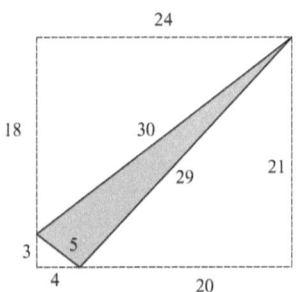

Figure PCK11

(c) Can you find three lattice points that are the vertices of an equilateral triangle? What about regular pentagon? Regular hexagon? Regular heptagon? Regular octagon?

(d) Consider question (c) for the three-dimensional case.

(e) Is there a theorem similar to the Pick theorem for the one-dimensional case? How about the three-dimensional case?

17.2 Lattice Equilateral Triangle

Let us consider the problem mentioned at the end of the previous section. That is, is there a lattice equilateral triangle? We shall establish the non-existence of a lattice equilateral triangle in a plane by computing its area in two ways: One asserts that the area of a lattice equilateral triangle (if it exists) must be an irrational number, while the other asserts that the area of an arbitrary lattice polygon (in particular, a lattice triangle, equilateral or not) must be rational. The contradiction establishes its non-existence.

The second assertion follows from the observation in the proof of the lemma in the previous section (Figure PCK8, page 131). The area of any lattice triangle (hence any lattice polygon) is either an integer or half of an integer (in either case, it must be a rational number).

On the other hand, let the length of a side of the lattice equilateral triangle (assuming it exists) be s. Then, by the Pythagorean theorem (page 11), the altitude h of the equilateral triangle (with side length s) is given by (Figure PCK12)

$$s^2 = h^2 + \left(\frac{s}{2}\right)^2, \quad h^2 = s^2 - \frac{1}{4}s^2, \quad h = \frac{\sqrt{3}}{2}s.$$

Therefore the area of the equilateral triangle is

$$\frac{1}{2} \cdot s \cdot h = \frac{\sqrt{3}}{4}s^2.$$

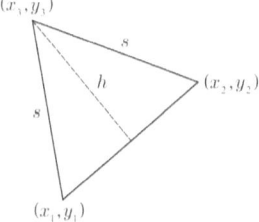

Figure PCK12

But

$$s^2 = (x_2 - x_1)^2 + (y_2 - y_1)^2,$$

where (x_1, y_1), (x_2, y_2) are the vertices of the triangle, and so s^2 is an integer (though s itself may be irrational), implying that the area of a lattice equilateral triangle (if it exists) must be irrational. The contradiction establishes that a lattice equilateral triangle can not exist.

We now give an alternate proof of the non-existence of a lattice equilateral triangle in a plane that does not depend on the irrationality of $\sqrt{3}$. Assume that a lattice equilateral triangle exists. Let its three vertices be (x_1, y_1), (x_2, y_2), (x_3, y_3), and

$$a_1 = x_2 - x_3, \quad a_2 = x_3 - x_1, \quad a_3 = x_1 - x_2,$$
$$b_1 = y_2 - y_3, \quad b_2 = y_3 - y_1, \quad b_3 = y_1 - y_2.$$

Then, clearly, we have

$$a_1 + a_2 + a_3 = 0, \quad b_1 + b_2 + b_3 = 0.$$

Furthermore, because the triangle is equilateral, we have

$$a_1^2 + b_1^2 = a_2^2 + b_2^2 = a_3^2 + b_3^2 \, (= K).$$

Now, being the sum of two perfect squares, K gives remainder 0, 1, or 2 (but never 3) when divided by 4. (Why?) Note that we do not consider the case $K = 0$, in which case the triangle degenerates to a point.

We start from the easiest case that K gives remainder 2 when divided by 4. In this case all a's (and all b's) must be odd integers. (Why?) But the sum of three odd integers can never be zero, which contradicts our assumption $a_1 + a_2 + a_3 = 0$. So this case can never happen.

Next, suppose K gives remainder 1 when divided by 4. In this case one of a_1 and b_1 is odd and the other even, hence their sum $(a_1 + b_1)$ must be odd. Same is true for $(a_2 + b_2)$ and $(a_3 + b_3)$. Hence the sum of three odd numbers $(a_1 + b_1) + (a_2 + b_2) + (a_3 + b_3)$ can not be zero. But, from the assumption,

$$(a_1 + b_1) + (a_2 + b_2) + (a_3 + b_3) = (a_1 + a_2 + a_3) + (b_1 + b_2 + b_3) = 0 + 0 = 0.$$

Again we obtain a contradiction. So this case can not happen either.

Finally, if K gives a remainder 0 when divided by 4, then all a_1, b_1, a_2, b_2, a_3, b_3, must be even. So we may replace each of them by its half. Denote their new replacements (note that they are still all integers) by the same notations. Then we still have

$$a_1 + a_2 + a_3 = 0, \quad b_1 + b_2 + b_3 = 0, \quad a_1^2 + b_1^2 = a_2^2 + b_2^2 = a_3^2 + b_3^2 \, (= K).$$

If this new K still gives 0 when divided by 4, then simply repeat the same procedure. Because $K \neq 0$, eventually, within a finite number of repetitions, we obtain a case that the newest K does not give remainder 0 when divided by 4; i.e., we eventually must reach one of the impossible cases discussed above. So in all cases, we get a contradiction, proving that a lattice equilateral triangle does not exist.

Note carefully, our proof is valid for any odd number of sides.

Theorem. In a plane, no lattice equilateral polygon of n sides exists unless n is even.

Exercise. Show that, for any even number $n \geq 4$, there exists a lattice equilateral n-gon.

17.3 Lattice Equiangular Polygons

We now investigate the lattice equiangular polygons in a plane. We shall not use any result from our discussion of lattice equilateral polygons. We start with examples. For any pair of integers p and q, not both zero, the points

$$(p, q), \quad (-q, p), \quad (-p, -q), \quad (q, -p)$$

are the vertices of an equiangular quadrangle (in fact, a square). (Figure PCK13) In addition, if $p > q > 0$, then

$$(p, q), (q, p), (-q, p), (-p, q), (-p, -q), (-q, -p), (q, -p), (p, -q)$$

are the vertices of an equiangular octagon. (Can you choose integers p and q such that this octagon becomes regular?)

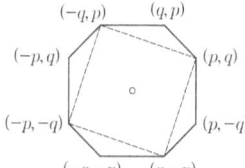

Figure PCK13

Exercise. Show that there are infinitely many lattice equiangular polygons no two of them are similar.

Because the sum of the exterior angles of any (convex) polygon is 2π, we see that an exterior angle φ of an equiangular m-gon (i.e., an equiangular polygon with m sides) is $2\pi/m$; in particular, φ/π is rational.

Now, following Pólya and Szegő,[1] we call a straight line that passes through (at least) two distinct lattice points a *lattice line*. Clearly, the slope of a lattice line is a rational number. Hence the value of the tangent function of the angle between two lattice lines must be rational because

$$\tan(x - y) = \frac{\tan x - \tan y}{1 + \tan x \cdot \tan y}.$$

But then so is the value of

$$\cos 2\varphi = \frac{\cos^2 \varphi - \sin^2 \varphi}{\cos^2 \varphi + \sin^2 \varphi} = \frac{1 - \tan^2 \varphi}{1 + \tan^2 \varphi}.$$

Thus we have the situation:

---

[1] *Problems and Theorems in Analysis*, Volume II (Springer-Verlag, Berlin, Heidelberg, New York, 1976), page 354.

Both φ/π and $\cos 2\varphi$ are rational.

We now exploit this situation. We start with the following.

> **Lemma.** A rational root of a monic polynomial with integer coefficients is an integer.

Recall that a polynomial is called *monic* if the leading coefficient is 1.

Proof. Suppose $\frac{p}{q}$ (where p and q are relatively prime integers and $q > 0$) is a root of a monic polynomial with integer coefficients:
$$x^n + a_1 x^{n-1} + a_2 x^{n-2} + \cdots + a_{n-1} x + a_n = 0;$$
i.e.,
$$\left(\frac{p}{q}\right)^n + a_1 \left(\frac{p}{q}\right)^{n-1} + a_2 \left(\frac{p}{q}\right)^{n-2} + \cdots + a_{n-1} \left(\frac{p}{q}\right) + a_n = 0.$$
Multiply both sides by q^{n-1}, we obtain
$$\frac{p^n}{q} + a_1 p^{n-1} + a_2 p^{n-2} q + \cdots + a_{n-1} p q^{n-2} + a_n q^{n-1} = 0,$$
where all the terms except the first are integers. Hence the first term must also be an integer. But p and q are relatively prime. Hence we must have $q = 1$; i.e., the root $\frac{p}{q}$ is an integer. □

The key to the next step is the following.

> **Lemma.** For each positive integer n, there exists a monic polynomial $f_n(x)$ of degree n with integer coefficients such that
> $$2\cos n\theta = f_n(x) \quad (x = 2\cos\theta).$$

Proof. The cases $n = 1, 2$, are trivial; simply let $f_1(x) = x$, and
$$f_2(x) = 2\cos 2\theta = 2(2\cos^2\theta - 1) = (2\cos\theta)^2 - 2 = x^2 - 2.$$
For the general case, it is easiest to appeal to complex numbers.[2] Let $z = \cos\theta + i\sin\theta$. Then, we have
$$z^n = \cos n\theta + i\sin n\theta \quad \text{and} \quad 2\cos n\theta = z^n + \frac{1}{z^n} \quad (n \geq 0).$$
And from the identity
$$\left(z^n + \frac{1}{z^n}\right)\left(z + \frac{1}{z}\right) = \left(z^{n+1} + \frac{1}{z^{n+1}}\right) + \left(z^{n-1} + \frac{1}{z^{n-1}}\right),$$

---

[2] See my book, *Complex Numbers and Geometry* (Mathematical Association of America, 1994).

Lattice Points

we obtain the recursive formula

$$f_{n+1}(x) = x f_n(x) - f_{n-1}(x).$$

Now the mathematical induction gives the desired conclusion in our lemma. □

Note that the proof above can be rewritten (at the expense of transparency[3]) so as the complex numbers do not appear.

Exercise. Prove that, in $f_n(x)$,

(i) all the coefficients of x^k, where k and n are of different parities, vanish;

(ii) the coefficient of x^{n-2} is $-n$;

(iii) $f_n(2) = 2$.

Now, if $\frac{\theta}{\pi}$ is rational, then for some integer n, $n\theta$ is a multiple of π, and so $\cos n\theta$ is an integer. But then $x = 2\cos\theta$ is a root of

$$2\cos n\theta = f_n(x),$$

which is a monic polynomial with integer coefficients (by lemma on page 136). Hence if $\cos\theta$ is also rational, then $x = 2\cos\theta$ is a rational root, and so, by lemma (page 136), must be an integer. But

$$|2\cos\theta| \leq 2,$$

hence we obtain the following.

Theorem. If both $\frac{\theta}{\pi}$ and $\cos\theta$ are rational, then the only possible values of $\cos\theta$ are 1, $\frac{1}{2}$, 0, $-\frac{1}{2}$, -1.

Applying the theorem to the exterior angle φ ($0 < \varphi \leq \frac{2\pi}{3}$), we obtain the following five cases:

1. $\cos 2\varphi = 1$. Then $\varphi = 0$. But an exterior angle can not be 0. So this case can not happen.

2. $\cos 2\varphi = \frac{1}{2}$. Then

$$\tan^2\varphi = \frac{1 - \cos 2\varphi}{1 + \cos 2\varphi} = \frac{1}{3}. \quad \text{But then} \quad \tan\varphi = \pm\frac{1}{\sqrt{3}} \quad \text{is not rational.}$$

So this case can not happen either.

---
[3] The shortest path between two truths in the real domain passes through the complex domain.—J. Hadamard

3. $\cos 2\varphi = 0$. Then

$$\tan^2 \varphi = \frac{1 - \cos 2\varphi}{1 + \cos 2\varphi} = 1, \quad \text{and} \quad \varphi = \frac{\pi}{4}.$$

Hence the equilateral polygon is an octagon. (Note that $\varphi \neq \frac{3\pi}{4}$. Why?)

4. $\cos 2\varphi = -\frac{1}{2}$. Then

$$\tan^2 \varphi = \frac{1 - \cos 2\varphi}{1 + \cos 2\varphi} = 3, \quad \text{and} \quad \tan \varphi = \pm\sqrt{3} \quad \text{is not rational.}$$

So this case can not happen either.

5. $\cos 2\varphi = -1$. Then $\varphi = \frac{\pi}{2}$, and the equiangular polygon is a rectangle.

We have shown:

> *Theorem.* The only lattice equiangular polygons in a plane are rectangles and octagons.

17.4 Lattice Regular Polygons

Finally, let us consider lattice regular polygons. We know lattice squares exist (in abundance). But how about a lattice regular octagon? Note that three consecutive vertices of a regular n-gon (a regular polygon with n sides) are the vertices of an isosceles triangle with side lengths a, a, b, say, and by the law of cosines, we have

$$b^2 = 2a^2 - 2a^2 \cos \frac{(n-2)\pi}{n}.$$

If the vertices are lattice points, then a^2 and b^2 are integers, implying that $\cos \frac{(n-2)\pi}{n}$ is rational. But for $n = 8$,

$$\cos \frac{6\pi}{8} = \cos \frac{3\pi}{4} = -\frac{1}{\sqrt{2}} \quad \text{is not rational.}$$

Hence a lattice regular octagon does not exist. We conclude:

> *Theorem.* The only lattice regular polygons in a plane are squares.

Exercise. In the 3-dimensional space,

(a) Give examples of a lattice equilateral triangle and a lattice regular hexagon.

(b) Show that the only lattice regular n-gons are $n = 3, 4, 6$.

Chapter 18

The Sums of Special Series

18.1 The Sum of the Powers

How do we find the sum of the first n natural numbers? That is, we want to find the sum
$$S_1 = 1 + 2 + 3 + \cdots + (n-1) + n.$$
This is a very typical arithmetic series. In a famous anecdote, the great mathematician Carl Friedrich Gauss (1777 - 1855) noticed, when he was a little school boy, the sum of the first term and the last, that of the second and the second from the last, the third and the third from the last, etc. are all the same. Therefore, we write the sum twice, second one in a reverse order:
$$\begin{aligned} S_1 &= 1 + 2 + 3 + \cdots + (n-1) + n \\ S_1 &= n + (n-1) + (n-2) + \cdots + 2 + 1 \end{aligned}$$
And adding the two equalities columnwise, we obtain
$$2S_1 = (n+1) + (n+1) + (n+1) + \cdots + (n+1).$$
Clearly, there are n $(n+1)$'s. So we have
$$2S_1 = n(n+1), \quad S_1 = \frac{n(n+1)}{2}.$$

So far so good. In fact, this trick works for all the arithmetic series. But how do we find the sum of the first n squares? That is, we want to find the value of
$$S_2 = 1^2 + 2^2 + 3^2 + \cdots + (n-1)^2 + n^2.$$
Let us try the same trick for this case; i.e., rewrite the series twice, with one of them in reverse order:
$$\begin{aligned} S_2 &= 1^2 + 2^2 + 3^2 + \cdots + (n-1)^2 + n^2 \\ S_2 &= n^2 + (n-1)^2 + (n-2)^2 + \cdots + 2^2 + 1^2 \end{aligned}$$

Now instead of the general case, let us check for an easy case, say, $n = 10$. Then the sum of the first and the last is $1^2 + 10^2 = 101$. But the sum of the second and the second from the last is $2^2 + 9^2 = 85$. That kills our method right there. This problem appears to be non-trivial. Luckily, there is a brilliant solution. It is simple to verify the identity
$$(k+1)^3 - (k-1)^3 = 6k^2 + 2.$$
Substituting $k = 1, 2, 3, \cdots, n$, successively, we obtain

$$\begin{aligned}
2^3 - 0^3 &= 6 \cdot 1^2 + 2 \\
3^3 - 1^3 &= 6 \cdot 2^2 + 2 \\
4^3 - 2^3 &= 6 \cdot 3^2 + 2 \\
5^3 - 3^3 &= 6 \cdot 4^2 + 2 \\
&\cdots \\
(n-1)^3 - (n-3)^3 &= 6 \cdot (n-2)^2 + 2 \\
n^3 - (n-2)^3 &= 6 \cdot (n-1)^2 + 2 \\
(n+1)^3 - (n-1)^3 &= 6 \cdot n^2 + 2
\end{aligned}$$

Then what? Add them up! The left-hand side gives the so-called "telescoping sum".
$$(n+1)^3 + n^3 - 0^3 - 1^3 = 6\left\{1^2 + 2^2 + 3^2 + \cdots + (n-1)^2 + n^2\right\} + 2n.$$
Therefore,
$$\begin{aligned}
6S_2 &= 2n^3 + 3n^2 + 3n - 2n = n\left(2n^2 + 3n + 1\right), \\
S_2 &= \frac{n(n+1)(2n+1)}{6}.
\end{aligned}$$

This is very nice. But how would anyone think of using the identity
$$(k+1)^3 - (k-1)^3 = 6k^2 + 2?$$

We postpone this question for a moment, and instead, we ask, "Would this method work in other cases?" Let us try it for
$$S_1 = 1 + 2 + 3 + \cdots + (n-1) + n.$$

We already know how to sum this one, but let us try our new method any way. This time we have
$$(k+1)^2 - (k-1)^2 = 4k.$$
Substituting $k = 1, 2, 3, \cdots, n$, successively, we obtain

$$\begin{aligned}
2^2 - 0^2 &= 4 \cdot 1 \\
3^2 - 1^2 &= 4 \cdot 2
\end{aligned}$$

The Sums of Special Series

$$4^2 - 2^2 = 4 \cdot 3$$
$$5^2 - 3^2 = 4 \cdot 4$$
$$\cdots \quad \cdots$$
$$(n-1)^2 - (n-3)^2 = 4 \cdot (n-2)$$
$$n^2 - (n-2)^2 = 4 \cdot (n-1)$$
$$(n+1)^2 - (n-1)^2 = 4 \cdot n$$

Again we have a telescoping sum on the left-hand side.

$$(n+1)^2 + n^2 - 0^2 - 1^2 = 4\{1 + 2 + 3 + \cdots + (n-1) + n\}.$$

Therefore,
$$4S_1 = 2n^2 + 2n = 2n(n+1), \quad S_1 = \frac{n(n+1)}{2},$$
in agreement with the result we obtained before.

Exercise. Find
$$S_3 = 1^3 + 2^3 + 3^3 + \cdots + (n-1)^3 + n^3.$$
Be sure to check your formula for the first few positive integers.

18.2 The Binomial Coefficients

Recall that the binomial coefficients are
$$\binom{n}{k} = \frac{n!}{k!(n-k)!} = \frac{n(n-1)(n-2)\cdots(n-k+1)}{k(k-1)\cdots 2 \cdot 1},$$
for nonnegative integers n and k, where $n!$ is the product of all the positive integers not greater than n; i.e., $n! = n \cdot (n-1) \cdots 2 \cdot 1$, and is called the *factorial* of n. Note that $\binom{n}{k} = 0$ if $n < k$. (Convention: $0! = 1$.)

Recall, from the first section,
$$1 + 2 + 3 + \cdots + (n-1) + n = \frac{n(n+1)}{2}.$$

It so happens that the right-hand side is $\binom{n+1}{2}$; i.e.,
$$S_1 = 1 + 2 + 3 + \cdots + (n-1) + n = \binom{n+1}{2}.$$

Hence it is only natural that we have
$$\binom{n+1}{2} - \binom{n}{2} = n.$$

Naturally, we can verify this identity directly. But once we have this identity, then we can substitute 1, 2, 3, $\cdots$, successively for n to obtain

$$\binom{2}{2} - \binom{1}{2} = 1$$

$$\binom{3}{2} - \binom{2}{2} = 2$$

$$\binom{4}{2} - \binom{3}{2} = 3$$

$$\binom{5}{2} - \binom{4}{2} = 4$$

$$\cdots \quad \cdots \quad \cdots$$

$$\binom{n}{2} - \binom{n-1}{2} = n-1$$

$$\binom{n+1}{2} - \binom{n}{2} = n$$

As before, we add them up. The left-hand side is a telescoping sum, obtaining

$$\binom{n+1}{2} - \binom{1}{2} = 1 + 2 + 3 + \cdots + (n-1) + n.$$

Therefore,

$$S_1 = \binom{n+1}{2} = \frac{n(n+1)}{2}.$$

Of course this is a circular argument. But, by hook or by crook, if we have

$$\binom{k+1}{2} - \binom{k}{2} = k,$$

then we can find the sum S_1 of the first n natural numbers.

Let us explore whether this method can be applied to other cases. For which we calculate a similar difference: $\binom{k+2}{3} - \binom{k}{3}$. Lo and behold, we have

$$\begin{aligned}\binom{k+2}{3} - \binom{k}{3} &= \frac{(k+2)(k+1)k}{3!} - \frac{k(k-1)(k-2)}{3!} \\ &= \frac{k}{6}\{(k+2)(k+1) - (k-1)(k-2)\} \\ &= \frac{k}{6}\{(k^2 + 3k + 2) - (k^2 - 3k + 2)\} \\ &= \frac{k}{6}(6k) = k^2.\end{aligned}$$

Once we have this identity, we are in good shape! Substituting $k = 1, 2, 3, \cdots, n$, successively, we obtain

$$\binom{3}{3} - \binom{1}{3} = 1^2$$

The Sums of Special Series

$$\binom{4}{3} - \binom{2}{3} = 2^2$$

$$\binom{5}{3} - \binom{3}{3} = 3^2$$

$$\binom{6}{3} - \binom{4}{3} = 4^2$$

$$\cdots \qquad \cdots \qquad \cdots$$

$$\binom{n}{3} - \binom{n-2}{3} = (n-2)^2$$

$$\binom{n+1}{3} - \binom{n-1}{3} = (n-1)^2$$

$$\binom{n+2}{3} - \binom{n}{3} = n^2$$

Again, we have a telescoping sum on the left-hand side, and we obtain

$$\binom{n+2}{3} + \binom{n+1}{3} - \binom{1}{3} - \binom{2}{3} = \sum_{k=1}^{n} k^2.$$

Therefore, we have

$$\begin{aligned} S_2 &= \frac{(n+2)(n+1)n}{3!} + \frac{(n+1)n(n-1)}{3!} \\ &= \frac{(n+1)n}{6}\{(n+2) + (n-1)\} \\ &= \frac{n(n+1)(2n+1)}{6}. \end{aligned}$$

Exercise. Can this method be extended further?

18.3 Faulhaber Polynomials

We saw in the previous sections that the sum S_1 of the first n positive integers is a quadratic polynomial in n. If we complete the square, we obtain

$$S_1 = \frac{1}{2}n(n+1) = \frac{1}{2}\left(n + \frac{1}{2}\right)^2 - \frac{1}{8}.$$

Thus considering n as a continuous variable in this formula, the graph is a parabola with (vertical) axis $n + \frac{1}{2} = 0$. In particular, the sum of the first n positive integers is an even quadratic polynomial in $(n + \frac{1}{2})$. Similarly,

$$S_2 = \frac{1}{6}n(n+1)(2n+1) = \frac{1}{3}\left(n + \frac{1}{2}\right)^3 - \frac{1}{12}\left(n + \frac{1}{2}\right).$$

The sum of the squares of the first n positive integers is an odd cubic polynomial in $(n + \frac{1}{2})$. It makes us wonder if similar relations holds for the sums of other powers? So let us recalculate from the very beginning. We have
$$\left(k + \frac{1}{2}\right)^2 - \left(k - \frac{1}{2}\right)^2 = 2k.$$
Substituting $k = 1, 2, 3, \cdots, n$, successively, we obtain
$$\left(\frac{3}{2}\right)^2 - \left(\frac{1}{2}\right)^2 = 2 \cdot 1,$$
$$\left(\frac{5}{2}\right)^2 - \left(\frac{3}{2}\right)^2 = 2 \cdot 2,$$
$$\left(\frac{7}{2}\right)^2 - \left(\frac{5}{2}\right)^2 = 2 \cdot 3,$$
$$\cdots \quad \cdots \quad \cdots$$
$$\left(n - \frac{1}{2}\right)^2 - \left(n - \frac{3}{2}\right)^2 = 2(n-1),$$
$$\left(n + \frac{1}{2}\right)^2 - \left(n - \frac{1}{2}\right)^2 = 2n.$$
Then adding these n equalities, we obtain
$$\left(n + \frac{1}{2}\right)^2 - \left(\frac{1}{2}\right)^2 = 2 \cdot (1 + 2 + 3 + \cdots + n) = 2S_1.$$
Therefore,
$$S_1 = \frac{1}{2}\left\{\left(n + \frac{1}{2}\right)^2 - \left(\frac{1}{2}\right)^2\right\}.$$
Of course we know this result already. Note that
$$\left(n + \frac{1}{2}\right)^2 = 2S_1 + \frac{1}{2^2}.$$
We shall use this equality repeatedly. For a reason that should be clear later, we proceed to compute the sums of the odd powers first.

We now compute the sum S_3 of the cubes of the first n positive integers. From the binomial theorem, we have
$$\left(k + \frac{1}{2}\right)^4 - \left(k - \frac{1}{2}\right)^4 = 2\left\{\binom{4}{1}\frac{k^3}{2} + \binom{4}{3}\frac{k}{2^3}\right\} = 4k^3 + k.$$
Substituting $k = 1, 2, 3, \cdots, n$, successively, and then adding the resulting equalities, we obtain
$$\left(n + \frac{1}{2}\right)^4 - \left(\frac{1}{2}\right)^4 = 4 \cdot (1^3 + 2^3 + 3^3 + \cdots + n^3) + (1 + 2 + 3 + \cdots + n)$$
$$= 4S_3 + S_1.$$

The Sums of Special Series

Because we know that S_1 is an even quadratic polynomial in $(n+\frac{1}{2})$, this equality implies that S_3 is an even quartic polynomial in $(n+\frac{1}{2})$. Furthermore, we have

$$
\begin{aligned}
S_3 &= \frac{1}{4}\left\{\left(n+\frac{1}{2}\right)^4 - \left(\frac{1}{2}\right)^4\right\} - \frac{1}{4}S_1 \\
&= \frac{1}{2}\left\{\left(n+\frac{1}{2}\right)^2 - \frac{1}{2^2}\right\} \cdot \frac{1}{2}\left\{\left(n+\frac{1}{2}\right)^2 + \frac{1}{2^2}\right\} - \frac{1}{2^2}S_1 \\
&= S_1 \cdot \frac{1}{2}\left\{\left(n+\frac{1}{2}\right)^2 + \frac{1}{2^2} - \frac{1}{2}\right\} \\
&= S_1 \cdot \frac{1}{2}\left\{\left(n+\frac{1}{2}\right)^2 - \frac{1}{2^2}\right\} = S_1^2.
\end{aligned}
$$

This is a surprising equality. It says, for any positive integer n, we have

$$1^3 + 2^3 + 3^3 + \cdots + n^3 = (1 + 2 + 3 + \cdots + n)^2.$$

Next, we compute the sum S_5 of the fifth powers of the first n positive integers. Again from the binomial theorem, we have

$$
\begin{aligned}
\left(k+\frac{1}{2}\right)^6 - \left(k-\frac{1}{2}\right)^6 &= 2\left\{\binom{6}{1}\frac{k^5}{2} + \binom{6}{3}\frac{k^3}{2^3} + \binom{6}{5}\frac{k}{2^5}\right\} \\
&= 6k^5 + 5k^3 + \frac{3}{8}k.
\end{aligned}
$$

It follows that

$$\left(n+\frac{1}{2}\right)^6 - \left(\frac{1}{2}\right)^6 = 6S_5 + 5S_3 + \frac{3}{8}S_1.$$

Again from our knowledge of S_1 and S_3, this equality implies that S_5 is an even polynomial of degree 6 in $(n+\frac{1}{2})$. Furthermore, solving for S_5, we obtain

$$
\begin{aligned}
S_5 &= \frac{1}{6}\left\{\left(n+\frac{1}{2}\right)^6 - \left(\frac{1}{2}\right)^6\right\} - \frac{1}{6}\left(5S_3 + \frac{3}{8}S_1\right) \\
&= \frac{1}{2}\left\{\left(n+\frac{1}{2}\right)^2 - \frac{1}{2^2}\right\} \cdot \frac{1}{3}\left\{\left(n+\frac{1}{2}\right)^4 + \left(n+\frac{1}{2}\right)^2\frac{1}{2^2} + \frac{1}{2^4}\right\} \\
&\quad - \frac{1}{6}\left(5S_1^2 + \frac{3}{2^3}S_1\right) \\
&= S_1 \cdot \frac{1}{3}\left\{\left(2S_1 + \frac{1}{2^2}\right)^2 + \frac{1}{2^2}\left(2S_1 + \frac{1}{2^2}\right) + \frac{1}{2^4}\right\} - \frac{S_1}{6}\left(5S_1 + \frac{3}{2^3}\right) \\
&= S_1 \cdot \frac{1}{3}\left\{\left(4S_1^2 + S_1 + \frac{1}{2^4}\right) + \frac{1}{2^2}\left(2S_1 + \frac{1}{2^2}\right) + \frac{1}{2^4}\right\} - \frac{1}{3}S_1\left(\frac{5}{2}S_1 + \frac{3}{2^4}\right) \\
&= \frac{S_1}{3}(4S_1^2 - S_1) = \frac{S_1^2}{3}(4S_1 - 1).
\end{aligned}
$$

It seems that we are beginning to see a pattern. Let us compute one more. Clearly, we have

$$\left(k+\frac{1}{2}\right)^8 - \left(k-\frac{1}{2}\right)^8 = 2\left\{\binom{8}{1}\frac{k^7}{2} + \binom{8}{3}\frac{k^5}{2^3} + \binom{8}{5}\frac{k^3}{2^5} + \binom{8}{7}\frac{k}{2^7}\right\}$$

$$= 8k^7 + 14k^5 + \frac{7}{2}k^3 + \frac{1}{8}k.$$

Therefore, we obtain

$$\left(n+\frac{1}{2}\right)^8 - \left(\frac{1}{2}\right)^8 = 8S_7 + 14S_5 + \frac{7}{2}S_3 + \frac{1}{8}S_1.$$

Again this equality implies that S_7 must be an even polynomial of degree 8 in $(n+\frac{1}{2})$. And we have

$$\begin{aligned}
S_7 &= \frac{1}{8}\left\{\left(n+\frac{1}{2}\right)^8 - \left(\frac{1}{2}\right)^8\right\} - \frac{1}{8}\left(14S_5 + \frac{7}{2}S_3 + \frac{1}{2^3}S_1\right) \\
&= \frac{1}{2}\left\{\left(n+\frac{1}{2}\right)^2 - \frac{1}{2^2}\right\} \cdot \frac{1}{4}\left\{\left(n+\frac{1}{2}\right)^6 + \left(n+\frac{1}{2}\right)^4\frac{1}{2^2} + \left(n+\frac{1}{2}\right)^2\frac{1}{2^4} + \frac{1}{2^6}\right\} \\
&\quad - \frac{1}{8}\left\{\frac{14S_1^2}{3}(4S_1-1) + \frac{7}{2}S_1^2 + \frac{1}{2^3}S_1\right\} \\
&= \frac{S_1}{4}\left\{\left(2S_1+\frac{1}{2^2}\right)^3 + \frac{1}{2^2}\left(2S_1+\frac{1}{2^2}\right)^2 + \frac{1}{2^4}\left(2S_1+\frac{1}{2^2}\right) + \frac{1}{2^6}\right\} \\
&\quad - \frac{S_1}{4}\left\{\frac{7}{3}(4S_1^2-S_1) + \frac{7}{2^2}S_1 + \frac{1}{2^4}\right\} \\
&= \frac{S_1}{4}\left\{\left(8S_1^3 + 3S_1^2 + \frac{3}{2^3}S_1 + \frac{1}{2^6}\right) + \frac{1}{2^2}\left(4S_1^2 + S_1 + \frac{1}{2^4}\right)\right. \\
&\quad \left. + \frac{1}{2^4}\left(2S_1+\frac{1}{2^2}\right) + \frac{1}{2^6}\right\} - \frac{S_1}{4}\left(\frac{28}{3}S_1^2 - \frac{7}{12}S_1 + \frac{1}{2^4}\right) \\
&= \frac{S_1^2}{4}\left(8S_1^2 - \frac{16}{3}S_1 + \frac{4}{3}\right) = \frac{S_1^2}{3}\left(6S_1^2 - 4S_1 + 1\right).
\end{aligned}$$

Having computed the sum S_p of the p-th powers of the first n positive integers; i.e.,

$$S_p = 1^p + 2^p + 3^p + \cdots + n^p,$$

for $p = 1, 3, 5, 7$, it should be an easy exercise to prove the following theorem (at least for the case that p is an odd positive integer).

> **Theorem.** For every odd positive integer p, S_p is an even polynomial of degree $p+1$ in $(n+\frac{1}{2})$. Furthermore, S_p can be expressed as the product of $S_1^2\,(=S_3)$ and a polynomial of degree $\frac{1}{2}(p-3)$ in S_1 if $p \geq 3$.

The Sums of Special Series

For every even positive integer p, S_p is an odd polynomial of degree $p+1$ in $(n+\frac{1}{2})$. Furthermore, S_p can be expressed as the product of S_2 and a polynomial of degree $\frac{1}{2}(p-2)$ in S_1.

We now embark on the case that p is even. For $p = 2$, from the binomial theorem, we have

$$\left(k+\frac{1}{2}\right)^3 - \left(k-\frac{1}{2}\right)^3 = 2\left\{\binom{3}{1}k^2\left(\frac{1}{2}\right) + \binom{3}{3}\left(\frac{1}{2}\right)^3\right\} = 3k^2 + \frac{1}{2^2}.$$

Substituting $k = 1, 2, 3, \cdots, n$, we obtain

$$\left(\frac{3}{2}\right)^3 - \left(\frac{1}{2}\right)^3 = 3 \cdot 1^2 + \frac{1}{2^2},$$

$$\left(\frac{5}{2}\right)^3 - \left(\frac{3}{2}\right)^3 = 3 \cdot 2^2 + \frac{1}{2^2},$$

$$\left(\frac{7}{2}\right)^3 - \left(\frac{5}{2}\right)^3 = 3 \cdot 3^2 + \frac{1}{2^2},$$

$$\cdots \qquad \cdots$$

$$\left(n-\frac{1}{2}\right)^3 - \left(n-\frac{3}{2}\right)^3 = 3(n-1)^2 + \frac{1}{2^2},$$

$$\left(n+\frac{1}{2}\right)^3 - \left(n-\frac{1}{2}\right)^3 = 3n^2 + \frac{1}{2^2}.$$

Then adding these n equalities, we obtain

$$\left(n+\frac{1}{2}\right)^3 - \left(\frac{1}{2}\right)^3 = 3 \cdot (1^2 + 2^2 + 3^2 + \cdots + n^2) + \frac{n}{2^2} = 3S_2 + \frac{n}{2^2}.$$

Combining the last terms of the two extreme ends, we see that S_2 is an odd polynomial in $(n+\frac{1}{2})$; in fact, we obtain

$$S_2 = \frac{1}{3}\left\{\left(n+\frac{1}{2}\right)^3 - \frac{1}{2^2}\left(n+\frac{1}{2}\right)\right\} = \frac{1}{3}\left(n+\frac{1}{2}\right)\left\{\left(n+\frac{1}{2}\right)^2 - \frac{1}{2^2}\right\}.$$

We now compute S_4. Again from the binomial theorem, we have

$$\left(k+\frac{1}{2}\right)^5 - \left(k-\frac{1}{2}\right)^5 = 2\left\{\binom{5}{1}\frac{k^4}{2} + \binom{5}{3}\frac{k^2}{2^3} + \binom{5}{5}\frac{1}{2^5}\right\}$$

$$= 5k^4 + \frac{5}{2}k^2 + \frac{1}{2^4}.$$

It follows that

$$\left(n+\frac{1}{2}\right)^5 - \left(\frac{1}{2}\right)^5 = 5S_4 + \frac{5}{2}S_2 + \frac{n}{2^4}.$$

Combining the last terms of both sides (as in the first line of the computation below), and knowing that S_2 is an odd polynomial in $(n+\frac{1}{2})$, we see that S_4 is also an odd polynomial in $(n+\frac{1}{2})$. Furthermore, as the last term on the right-hand side is moved to the left-hand side, then (as the third line in the computation below shows that) the new left-hand side contains factor S_2, and so S_4 must also contain factor S_2. In fact, we have

$$\begin{aligned}
S_4 &= \frac{1}{5}\left\{\left(n+\frac{1}{2}\right)^5 - \frac{1}{2^4}\left(n+\frac{1}{2}\right)\right\} - \frac{1}{2}S_2 \\
&= \frac{1}{5}\left(n+\frac{1}{2}\right)\left\{\left(n+\frac{1}{2}\right)^4 - \frac{1}{2^4}\right\} - \frac{1}{2}S_2 \\
&= \frac{1}{3}\left(n+\frac{1}{2}\right)\left\{\left(n+\frac{1}{2}\right)^2 - \frac{1}{2^2}\right\} \cdot \frac{3}{5}\left\{\left(n+\frac{1}{2}\right)^2 + \frac{1}{2^2}\right\} - \frac{1}{2}S_2 \\
&= S_2 \cdot \frac{1}{5}\left\{3\left(2S_1 + \frac{1}{2^2}\right) + \frac{3}{2^2} - \frac{5}{2}\right\} \\
&= \frac{S_2}{5}(6S_1 - 1).
\end{aligned}$$

Next we compute S_6. Again from the binomial theorem, we have

$$\begin{aligned}
\left(k+\frac{1}{2}\right)^7 - \left(k-\frac{1}{2}\right)^7 &= 2\left\{\binom{7}{1}\frac{k^6}{2} + \binom{7}{3}\frac{k^4}{2^3} + \binom{7}{5}\frac{k^2}{2^5} + \binom{7}{7}\frac{1}{2^7}\right\} \\
&= 7k^6 + \frac{35}{2^2}k^4 + \frac{21}{2^4}k^2 + \frac{1}{2^6}.
\end{aligned}$$

Therefore, we obtain

$$\left(n+\frac{1}{2}\right)^7 - \left(\frac{1}{2}\right)^7 = 7S_6 + \frac{35}{2^2}S_4 + \frac{21}{2^4}S_2 + \frac{n}{2^6}.$$

Again combining the last terms of both sides, and knowing that S_2 and S_4 are odd polynomials in $(n+\frac{1}{2})$, we see that S_6 is also an odd polynomial in $(n+\frac{1}{2})$. Furthermore, as the last term on the right-hand side is moved to the left-hand side, then (it is easy to see that) the new left-hand side contains factor S_2, and all the remaining terms on the right-hand side except S_6 contain factor S_2, hence S_6 must also contain factor S_2. In fact, we have

$$\begin{aligned}
S_6 &= \frac{1}{7}\left\{\left(n+\frac{1}{2}\right)^7 - \frac{1}{2^6}\left(n+\frac{1}{2}\right)\right\} - \frac{1}{2^2}\left(5S_4 + \frac{3}{4}S_2\right) \\
&= \frac{1}{3}\left(n+\frac{1}{2}\right)\left\{\left(n+\frac{1}{2}\right)^2 - \frac{1}{2^2}\right\} \cdot \frac{3}{7}\left\{\left(n+\frac{1}{2}\right)^4 + \left(n+\frac{1}{2}\right)^2\frac{1}{2^2} + \frac{1}{2^4}\right\} \\
&\quad - \frac{1}{4}\left\{S_2(6S_1 - 1) + \frac{3}{4}S_2\right\}
\end{aligned}$$

The Sums of Special Series

$$= \frac{S_2}{7} \cdot 3 \left\{ \left(2S_1 + \frac{1}{2^2}\right)^2 + \frac{1}{2^2}\left(2S_1 + \frac{1}{2^2}\right) + \frac{1}{2^4} \right\}$$

$$- \frac{S_2}{4} \left\{ (6S_1 - 1) + \frac{3}{2^2} \right\}$$

$$= \frac{S_2}{7} \left\{ \left(12S_1^2 + 3S_1 + \frac{3}{2^4}\right) + \left(\frac{3}{2}S_1 + \frac{3}{2^4}\right) + \frac{3}{2^4} \right\}$$

$$- \frac{S_2}{7} \left\{ \left(\frac{21}{2}S_1 - \frac{7}{2^2}\right) + \frac{21}{2^4} \right\}$$

$$= \frac{S_2}{7} \left(12S_1^2 - 6S_1 + 1 \right).$$

Now we are becoming familiar with this computation. Let us compute just one more quickly. We have

$$\left(k + \frac{1}{2}\right)^9 - \left(k - \frac{1}{2}\right)^9 = 2\left\{\binom{9}{1}\frac{k^8}{2} + \binom{9}{3}\frac{k^6}{2^3} + \binom{9}{5}\frac{k^4}{2^5} + \binom{9}{7}\frac{k^2}{2^7} + \binom{9}{9}\frac{1}{2^9}\right\}$$

$$= 9k^8 + 84 \cdot \frac{k^6}{2^2} + 126 \cdot \frac{k^4}{2^4} + 36 \cdot \frac{k^2}{2^6} + \frac{1}{2^8}.$$

$$\left(n + \frac{1}{2}\right)^9 - \left(\frac{1}{2}\right)^9 = 9S_8 + 21S_6 + \frac{63}{8}S_4 + \frac{9}{16}S_2 + \frac{n}{2^8}$$

$$\left(n + \frac{1}{2}\right)^9 - \frac{1}{2^8}\left(n + \frac{1}{2}\right) = 9S_8 + 21S_6 + \frac{63}{8}S_4 + \frac{9}{16}S_2.$$

$$S_8 = \frac{S_2}{3}\left\{\left(n + \frac{1}{2}\right)^6 + \left(n + \frac{1}{2}\right)^4 \cdot \frac{1}{2^2} + \left(n + \frac{1}{2}\right)^2 \cdot \frac{1}{2^4} + \frac{1}{2^6}\right\}$$

$$- \frac{1}{9}\left\{21S_6 + \frac{63}{8}S_4 + \frac{9}{16}S_2\right\}$$

$$= \frac{S_2}{3}\left\{\left(2S_1 + \frac{1}{2^2}\right)^3 + \frac{1}{2^2}\left(2S_1 + \frac{1}{2^2}\right)^2 + \frac{1}{2^4}\left(2S_1 + \frac{1}{2^2}\right) + \frac{1}{2^6}\right\}$$

$$- S_2 \left\{ \frac{7}{3} \cdot \frac{1}{7}(12S_1^2 - 6S_1 + 1) + \frac{7}{8} \cdot \frac{1}{5}(6S_1 - 1) + \frac{1}{16} \right\}$$

$$= \frac{S_2}{3}\left\{\left(8S_1^3 + 3S_1^2 + \frac{3}{8}S_1 + \frac{1}{2^6}\right)\right.$$

$$\left. + \frac{1}{2^2}\left(4S_1^2 + S_1 + \frac{1}{2^4}\right) + \frac{1}{2^4}\left(2S_1 + \frac{1}{2^2}\right) + \frac{1}{2^6}\right\}$$

$$- S_2 \left\{ \left(4S_1^2 - 2S_1 + \frac{1}{3}\right) + \left(\frac{21}{20}S_1 - \frac{7}{40}\right) + \frac{1}{16} \right\}$$

$$= \frac{S_2}{3}\left(8S_1^3 + 4S_1^2 + \frac{3}{4}S_1 + \frac{1}{2^4}\right) - S_2\left(4S_1^2 - \frac{19}{20}S_1 + \frac{53}{2^2 \cdot 3 \cdot 5}\right)$$

$$= \frac{S_2}{15}\left(40S_1^3 - 40S_1^2 + 18S_1 - 3\right).$$

Now enough information is provided to finish the proof of the theorem above. Note again that to prove the theorem, the computation need not be carried out all the way to the end.

Exercise.[1] For any positive integers p and n, set

$$S_p(n) = 1^p + 2^p + 3^p + \cdots + n^p.$$

Then $S_p(n)$ is a polynomial in n. Hence $S_p(n)$ is meaningful not just for positive integers n. Show that if p is an odd positive integer, then

$$S_p(-n) = S_p(n-1) \quad \text{for all integers } n;$$

while if p is an even positive integer, then

$$S_p(-n) = -S_p(n-1) \quad \text{for all integers } n.$$

Hint. Use the recursive formula

$$S_p(n+1) = S_p(n) + n^p,$$

and the mathematical induction.
Is the converse true? That is, suppose $f(x)$ is a function defined on all real numbers, and

$$f(-x) = f(x-1) \quad \text{for all real numbers } x.$$

Is it necessary that $f(x)$ is an even function of $t = x + \frac{1}{2}$?
What if $f(-x) = -f(x-1)$?

18.4 The Sums of the Reciprocals of $S_p(n)$

For positive integer p, set

$$S_p(k) = 1^p + 2^p + 3^p + \cdots + k^p.$$

Exercise. Show that

(a) $\displaystyle\sum_{k=1}^{\infty} \frac{1}{S_1(k)} = 2;$

(b) $\displaystyle\sum_{k=1}^{\infty} \frac{1}{S_3(k)} = \frac{4}{3}\left(\pi^2 - 9\right) = 1.1594\cdots$
  (assuming that[2] $\sum_{k=1}^{\infty} \frac{1}{k^2} = \pi^2/6$).

---
[1] The idea of considering $S_p(n)$ on the whole complex plane was suggested by Professor Reuben Hersh of University of New Mexico (retired).
[2] See Hahn and Epstein, *Classical Complex Analysis*, (Jones and Bartlett, 1996), page 215.

The Sums of Special Series

We now compute the case $p = 2$.

$$\sum_{k=1}^{\infty} \frac{1}{S_2(k)} = \sum_{k=1}^{\infty} \frac{6}{k(k+1)(2k+1)} = 6\sum_{k=1}^{\infty} \left(\frac{1}{k} + \frac{1}{k+1} - \frac{4}{2k+1}\right).$$

Note carefully, although the series on the left-hand side converges absolutely, none of the three series on the right-hand side converges by itself.

$$\sum_{k=1}^{n} \left(\frac{1}{k} + \frac{1}{k+1} - \frac{4}{2k+1}\right) = 2\sum_{k=1}^{n} \left(\frac{1}{2k} - \frac{2}{2k+1} + \frac{1}{2k+2}\right)$$

$$= 2\left\{\sum_{k=1}^{n} \left(\frac{2}{2k} - \frac{2}{2k+1}\right) - \frac{1}{2} + \frac{1}{2n+2}\right\}$$

$$= 4\left\{\sum_{k=1}^{2n+1} \frac{(-1)^k}{k} + 1\right\} - 1 + \frac{1}{n+1}$$

$$= 4\sum_{k=1}^{2n+1} \frac{(-1)^k}{k} + 3 + \frac{1}{n+1}.$$

Clearly, the last term on the right-hand side tends to 0 (as $n \to \infty$). Furthermore,

$$\sum_{k=1}^{2n+1} \frac{(-1)^k}{k} = \sum_{k=1}^{2n+1} \int_0^1 (-1)^k x^{k-1}\, dx = \int_0^1 \left\{\sum_{k=1}^{2n+1} (-1)^k x^{k-1}\right\} dx$$

$$= \int_0^1 \frac{-1 + x^{2n+1}}{1+x} dx = -\int_0^1 \frac{dx}{1+x} + \int_0^1 \frac{x^{2n+1}}{1+x} dx.$$

$$\int_0^1 \frac{dx}{1+x} = [\log(1+x)]_0^1 = \log 2.$$

$$0 < \int_0^1 \frac{x^{2n+1}}{1+x} dx < \int_0^1 x^{2n+1} dx = \frac{1}{2n+2} \to 0 \quad (\text{as } n \to \infty).$$

Thus putting all our computations together, we obtain

$$\sum_{k=1}^{\infty} \frac{1}{S_2(k)} = 18 - 24\log 2 \ (= 1.3644\cdots).$$

The case $p = 4$ involves the gamma function and can not be expressed in terms of elementary functions.

Finally, we compute the case $p = 5$. We have

$$S_5(k) = \frac{1}{6}k^2(k+1)^2\left(k^2 + k - \frac{1}{2}\right),$$

$$\frac{1}{S_5(k)} = \frac{24}{k^2 + k - \frac{1}{2}} - \frac{24}{k(k+1)} - \frac{12}{k^2(k+1)^2},$$

$$\sum_{k=1}^{\infty} \frac{1}{S_5(k)} = 24\sum_{k=1}^{\infty} \frac{1}{k^2 + k - \frac{1}{2}} - 24\sum_{k=1}^{\infty} \frac{1}{k(k+1)} - 12\sum_{k=1}^{\infty} \frac{1}{k^2(k+1)^2}.$$

The values of the last two series are known from the exercise at the beginning of this section, so we have only to compute the sum of the first series. Set

$$k^2 + k - \frac{1}{2} = (k-a)(k-b),$$

where

$$a = \frac{-1+\sqrt{3}}{2}, \quad b = \frac{-1-\sqrt{3}}{2}.$$

Then

$$\frac{1}{k^2+k-\frac{1}{2}} = \frac{1}{a-b}\left(\frac{1}{k-a} - \frac{1}{k-b}\right) = \frac{1}{\sqrt{3}}\left(\frac{1}{k-a} - \frac{1}{k+a+1}\right)$$

$$= \frac{1}{\sqrt{3}}\left\{\left(\frac{1}{k-a} - \frac{1}{k+a}\right) + \left(\frac{1}{k+a} - \frac{1}{k+a+1}\right)\right\},$$

$$\sum_{k=1}^{\infty} \frac{1}{k^2+k-\frac{1}{2}} = \frac{1}{\sqrt{3}} \sum_{k=1}^{\infty}\left(\frac{1}{k-a} - \frac{1}{k+a}\right) + \frac{1}{\sqrt{3}} \sum_{k=1}^{\infty}\left(\frac{1}{k+a} - \frac{1}{k+a+1}\right).$$

Now, the partial fraction expansion of the cotangent function[3]

$$\pi \cot \pi z = \frac{1}{z} + \sum_{k=1}^{\infty}\left(\frac{1}{z-k} + \frac{1}{z+k}\right)$$

gives

$$\sum_{k=1}^{\infty}\left(\frac{1}{k-a} - \frac{1}{k+a}\right) = -\sum_{k=1}^{\infty}\left(\frac{1}{a-k} + \frac{1}{a+k}\right) = \frac{1}{a} - \pi \cot \pi a.$$

And

$$\sum_{k=1}^{\infty}\left(\frac{1}{k+a} - \frac{1}{k+a+1}\right) = \frac{1}{1+a} = -\frac{1}{b}.$$

Therefore,

$$\sum_{k=1}^{\infty} \frac{1}{S_5(k)} = \frac{24}{\sqrt{3}}\left(\frac{1}{a} - \pi \cot \pi a - \frac{1}{b}\right) - 24 - \frac{12}{3}\left(\pi^2 - 9\right)$$

$$= \frac{24}{\sqrt{3}}\left(2\sqrt{3} - \pi \cot \pi a\right) - 24 - 4\left(\pi^2 - 9\right)$$

$$= 60 - 4\pi^2 - 8\sqrt{3} \cot\left(\frac{\sqrt{3}-1}{2}\pi\right) \quad (= 1.035\cdots).$$

---

[3] See Hahn and Epstein, *Classical Complex Analysis* (Jones and Bartlett, 1996), page 214.

18.5 The Sums of Trigonometric Functions

We now investigate whether we could imitate the method in the first two sections for trigonometric series. Specifically, we want to prove the formulas

$$\sin\theta + \sin 2\theta + \sin 3\theta + \cdots + \sin n\theta = \sum_{k=1}^{n} \sin k\theta = \frac{\sin\frac{n+1}{2}\theta \cdot \sin\frac{n}{2}\theta}{\sin\frac{\theta}{2}},$$

$$\cos\theta + \cos 2\theta + \cos 3\theta + \cdots + \cos n\theta = \sum_{k=1}^{n} \cos k\theta = \frac{\cos\frac{n+1}{2}\theta \cdot \sin\frac{n}{2}\theta}{\sin\frac{\theta}{2}},$$

provided that θ is not a multiple of π. (See also page 222.) We start from the sine series. Recall that

$$\cos(k-1)\theta - \cos(k+1)\theta = 2\sin k\theta \cdot \sin\theta.$$

Substituting $k = 1, 2, 3, \cdots, n$, in this equality, we obtain

$$\begin{aligned}
1 - \cos 2\theta &= 2\sin\theta \cdot \sin\theta, \\
\cos\theta - \cos 3\theta &= 2\sin 2\theta \cdot \sin\theta, \\
\cos 2\theta - \cos 4\theta &= 2\sin 3\theta \cdot \sin\theta, \\
\cdots \quad &= \quad \cdots \\
\cos(n-3)\theta - \cos(n-1)\theta &= 2\sin(n-2)\theta \cdot \sin\theta, \\
\cos(n-2)\theta - \cos n\theta &= 2\sin(n-1)\theta \cdot \sin\theta, \\
\cos(n-1)\theta - \cos(n+1)\theta &= 2\sin n\theta \cdot \sin\theta.
\end{aligned}$$

Summing up these n equalities, we obtain

$$(1 + \cos\theta) - [\cos n\theta + \cos(n+1)\theta] = 2\sin\theta \cdot \sum_{k=1}^{n} \sin k\theta,$$

$$2\cos^2\frac{\theta}{2} - 2\cos\frac{2n+1}{2}\theta \cdot \cos\frac{\theta}{2} = 4\sin\frac{\theta}{2} \cdot \cos\frac{\theta}{2} \cdot \sum_{k=1}^{n} \sin k\theta.$$

Cancelling the common factor $2\cos\frac{\theta}{2}$ from both sides, we obtain

$$2\sin\frac{\theta}{2} \cdot \sum_{k=1}^{n} \sin k\theta = \cos\frac{\theta}{2} - \cos\frac{2n+1}{2}\theta = 2\sin\frac{n+1}{2}\theta \cdot \sin\frac{n}{2}\theta.$$

Therefore,

$$\sum_{k=1}^{n} \sin k\theta = \frac{\sin\frac{n+1}{2}\theta \cdot \sin\frac{n}{2}\theta}{\sin\frac{\theta}{2}}.$$

Note that in the computation above, we divided both side twice, the first time by $2\cos\frac{\theta}{2}$, and the second time by $\sin\frac{\theta}{2}$. Hence our result is valid only if $2\cos\frac{\theta}{2}\cdot\sin\frac{\theta}{2} = \sin\theta \neq 0$; i.e., θ is not a multiple of π.

The case for cosine series can be carried out in a similar way. Recall that

$$\sin(k+1)\theta - \sin(k-1)\theta = 2\cos k\theta \cdot \sin\theta.$$

Substituting $k = 1, 2, 3, \cdots, n$, in this equality, we obtain

$$\begin{aligned}
\sin 2\theta - 0 &= 2\cos\theta \cdot \sin\theta, \\
\sin 3\theta - \sin\theta &= 2\cos 2\theta \cdot \sin\theta, \\
\sin 4\theta - \sin 2\theta &= 2\cos 3\theta \cdot \sin\theta, \\
&\cdots \\
\sin(n-1)\theta - \sin(n-3)\theta &= 2\cos(n-2)\theta \cdot \sin\theta, \\
\sin n\theta - \sin(n-2)\theta &= 2\cos(n-1)\theta \cdot \sin\theta, \\
\sin(n+1)\theta - \sin(n-1)\theta &= 2\cos n\theta \cdot \sin\theta.
\end{aligned}$$

Again summing up these n equalities, we obtain

$$[\sin(n+1)\theta + \sin n\theta] - \sin\theta = 2\sin\theta \cdot \sum_{k=1}^{n}\cos k\theta,$$

$$2\sin\frac{2n+1}{2}\theta \cdot \cos\frac{\theta}{2} - 2\sin\frac{\theta}{2} \cdot \cos\frac{\theta}{2} = 4\sin\frac{\theta}{2} \cdot \cos\frac{\theta}{2} \cdot \sum_{k=1}^{n}\cos k\theta.$$

Cancelling the common factor $2\cos\frac{\theta}{2}$ from both sides, we obtain

$$2\sin\frac{\theta}{2} \cdot \sum_{k=1}^{n}\cos k\theta = \sin\frac{2n+1}{2}\theta - \sin\frac{\theta}{2} = 2\cos\frac{n+1}{2}\theta \cdot \sin\frac{n}{2}\theta.$$

Therefore,

$$\sum_{k=1}^{n}\cos k\theta = \frac{\cos\frac{n+1}{2}\theta \cdot \sin\frac{n}{2}\theta}{\sin\frac{\theta}{2}},$$

provided that θ is not a multiple of π.

We now present an alternate proof. Again we start from the sine series. This time we use the equality

$$\sin(k+1)\theta + \sin(k-1)\theta = 2\sin k\theta \cdot \cos\theta.$$

Substituting $k = 1, 2, 3, \cdots, n$, and summing up, we obtain

$$\sum_{k=1}^{n}\sin(k+1)\theta + \sum_{k=1}^{n}\sin(k-1)\theta = 2\cos\theta \cdot \sum_{k=1}^{n}\sin k\theta.$$

The Sums of Special Series

Obviously, we have

$$\sum_{k=1}^{n} \sin(k+1)\theta = \left\{\sum_{k=1}^{n} \sin k\theta\right\} + \sin(n+1)\theta - \sin\theta,$$

$$\sum_{k=1}^{n} \sin(k-1)\theta = \left\{\sum_{k=1}^{n} \sin k\theta\right\} - \sin n\theta.$$

Substituting these results into the equality above and collecting the like terms, we obtain

$$2(1-\cos\theta) \cdot \sum_{k=1}^{n} \sin k\theta = \sin\theta - [\sin(n+1)\theta - \sin n\theta],$$

$$4\sin^2 \frac{\theta}{2} \cdot \sum_{k=1}^{n} \sin k\theta = 2\sin\frac{\theta}{2} \cdot \cos\frac{\theta}{2} - 2\cos\frac{2n+1}{2}\theta \cdot \sin\frac{\theta}{2}.$$

Cancelling the common factor $2\sin\frac{\theta}{2}$ from both sides, we obtain

$$2\sin\frac{\theta}{2} \cdot \sum_{k=1}^{n} \sin k\theta = \cos\frac{\theta}{2} - \cos\frac{2n+1}{2}\theta = 2\sin\frac{n+1}{2}\theta \cdot \sin\frac{n}{2}\theta.$$

Therefore,

$$\sum_{k=1}^{n} \sin k\theta = \frac{\sin\frac{n+1}{2}\theta \cdot \sin\frac{n}{2}\theta}{\sin\frac{\theta}{2}}.$$

Finally, we compute the cosine series. This time, we use the equality

$$\cos(k+1)\theta + \cos(k-1)\theta = 2\cos k\theta \cdot \cos\theta.$$

Substituting $k = 1, 2, 3, \cdots, n$, and summing up, we obtain

$$\sum_{k=1}^{n} \cos(k+1)\theta + \sum_{k=1}^{n} \cos(k-1)\theta = 2\cos\theta \cdot \sum_{k=1}^{n} \cos k\theta.$$

Obviously, we have

$$\sum_{k=1}^{n} \cos(k+1)\theta = \left\{\sum_{k=1}^{n} \cos k\theta\right\} + \cos(n+1)\theta - \cos\theta,$$

$$\sum_{k=1}^{n} \cos(k-1)\theta = \left\{\sum_{k=1}^{n} \cos k\theta\right\} + 1 - \cos n\theta.$$

Substituting these results into the equality above and collecting the like terms, we obtain

$$2(1-\cos\theta) \cdot \sum_{k=1}^{n} \cos k\theta = [\cos n\theta - \cos(n+1)\theta] - (1-\cos\theta),$$

$$4\sin^2\frac{\theta}{2} \cdot \sum_{k=1}^{n} \cos k\theta = 2\sin\frac{2n+1}{2}\theta \cdot \sin\frac{\theta}{2} - 2\sin^2\frac{\theta}{2}.$$

Cancelling the common factor $2\sin\frac{\theta}{2}$ from both sides, we obtain

$$2\sin\frac{\theta}{2} \cdot \sum_{k=1}^{n} \cos k\theta = \sin\frac{2n+1}{2}\theta - \sin\frac{\theta}{2} = 2\cos\frac{n+1}{2}\theta \cdot \sin\frac{n}{2}\theta.$$

Therefore,

$$\sum_{k=1}^{n} \cos k\theta = \frac{\cos\frac{n+1}{2}\theta \cdot \sin\frac{n}{2}\theta}{\sin\frac{\theta}{2}}.$$

Exercise. Prove: If θ is not a multiple of π, then

$$\sin\varphi + \sin(\theta+\varphi) + \sin(2\theta+\varphi) + \cdots + \sin(n\theta+\varphi)$$
$$= \sum_{k=0}^{n} \sin(k\theta+\varphi) = \frac{\sin(\frac{n}{2}\theta+\varphi) \cdot \sin\frac{n+1}{2}\theta}{\sin\frac{\theta}{2}},$$
$$\cos\varphi + \cos(\theta+\varphi) + \cos(2\theta+\varphi) + \cdots + \cos(n\theta+\varphi)$$
$$= \sum_{k=0}^{n} \cos(k\theta+\varphi) = \frac{\cos(\frac{n}{2}\theta+\varphi) \cdot \sin\frac{n+1}{2}\theta}{\sin\frac{\theta}{2}}.$$

Hint. For the sine series, use one of the following.

$$\cos[(k-1)\theta+\varphi] - \cos[(k+1)\theta+\varphi] = 2\sin(k\theta+\varphi) \cdot \sin\theta,$$
$$\sin[(k-1)\theta+\varphi] + \sin[(k+1)\theta+\varphi] = 2\sin(k\theta+\varphi) \cdot \cos\theta.$$

And for the cosine series, use one of the following.

$$\sin[(k+1)\theta+\varphi] - \sin[(k-1)\theta+\varphi] = 2\cos(k\theta+\varphi) \cdot \sin\theta,$$
$$\cos[(k+1)\theta+\varphi] + \cos[(k-1)\theta+\varphi] = 2\cos(k\theta+\varphi) \cdot \cos\theta.$$

Chapter 19
The Morley Theorem

Because of the fact that there exist angles (such as 60° angle) that cannot be trisected (using compass and straightedge finitely many times), mathematicians tended to shy away from problems involving trisections of angles. Perhaps that is why the following dandy theorem was discovered only in 1904 by the British geometer Frank Morley (1860 - 1937). The idea in our proof was essentially the one shown to me by Professor John H. Conway of Princeton University, when I (as the director of the New Mexico Mathematics Contest) invited him to give a talk to the Contest finalists in February 1999.

Theorem [Morley] In any triangle, the three points of intersection of adjacent angle trisectors are the vertices of an equilateral triangle. (Figure MLY)

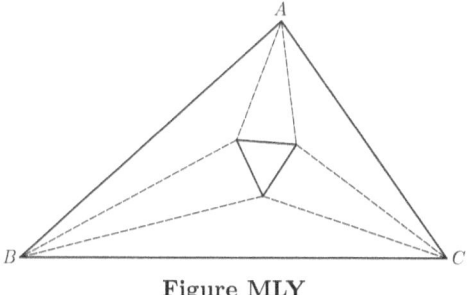

Figure MLY

Proof. Let the three angles of a given triangle ABC be 3α, 3β, 3γ. (Clearly, $\alpha + \beta + \gamma = \frac{\pi}{3}$ radian $(= 60°)$.) The theorem mentions only about the angles (of the given triangle), but nothing about its side lengths. Hence if the theorem is true for $\triangle LMN$ say, then it is also true for all the triangles that is similar to $\triangle LMN$. Thus we want to construct $\triangle LMN$ similar to the given triangle ABC, yet for which the theorem is easy to verify. We start by attaching triangles with angles

$$\{\alpha, \beta', \gamma'\}, \quad \{\beta, \gamma' \alpha'\}, \quad \{\gamma, \alpha', \beta'\},$$

where

$$\alpha' = \alpha + \frac{\pi}{3}, \quad \beta' = \beta + \frac{\pi}{3}, \quad \gamma' = \gamma + \frac{\pi}{3},$$

to the three sides of an arbitrary equilateral triangle UVW outwardly. (Figure MLY1) Note carefully, such triangles do exist; i.e., in each case the sum of the three angles is π radian $(= 180°)$. (Figure MLY1) To spell out in detail,

$$\angle WLV = \alpha, \quad \angle VWL = \beta', \quad \angle LVW = \gamma';$$
$$\angle UMW = \beta, \quad \angle WUM = \gamma', \quad \angle MWU = \alpha';$$
$$\angle VNU = \gamma, \quad \angle UVN = \alpha', \quad \angle NUV = \beta'.$$

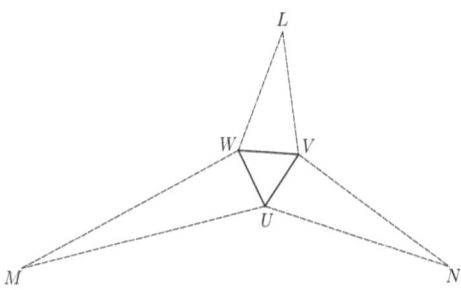

Figure MLY1

Let X be the intersection of MW and NV. (Figure MLY2) Note that they do intersect because

$$\angle MWV + \angle WVN = (\angle MWU + \angle UWV) + (\angle WVU + \angle UVN)$$
$$= 2\left(\alpha' + \frac{\pi}{3}\right) = 2\alpha + 4 \cdot \frac{\pi}{3} > \pi.$$

Then

$$\angle XWV = \pi - (\angle MWU + \angle UWV) = \pi - \left(\alpha' + \frac{\pi}{3}\right) = \frac{\pi}{3} - \alpha,$$
$$\angle XVW = \pi - (\angle NVU + \angle UVW) = \pi - \left(\alpha' + \frac{\pi}{3}\right) = \frac{\pi}{3} - \alpha.$$

And so $\triangle XWV$ is isosceles, implying that $\overline{XW} = \overline{XV}$. It follows that

$$\triangle XUW \cong \triangle XUV \quad (SSS).$$

In particular, $\angle UXM = \angle UXW = \angle UXV = \angle UXN$. Therefore, XU is the angle bisector of $\angle MXN$.

The Morley Theorem

Now
$$\angle MXN = \angle WXV = \pi - (\angle XWV + \angle XVW) = \pi - 2\left(\frac{\pi}{3} - \alpha\right) = \frac{\pi}{3} + 2\alpha,$$
and
$$\begin{aligned}
\angle MUN &= 2\pi - (\angle MUW + \angle WUV + \angle VUN) \\
&= 2\pi - (\gamma' + \frac{\pi}{3} + \beta') = 2\pi - \left[\left(\frac{\pi}{3} + \gamma\right) + \frac{\pi}{3} + \left(\frac{\pi}{3} + \beta\right)\right] \\
&= 2\pi - [\pi + (\alpha + \beta + \gamma) - \alpha] = \pi - \frac{\pi}{3} + \alpha = \frac{2\pi}{3} + \alpha.
\end{aligned}$$

It follows that
$$\frac{1}{2}(\angle MXN + \pi) = \frac{1}{2}\left(\frac{\pi}{3} + 2\alpha + \pi\right) = \frac{2\pi}{3} + \alpha = \angle MUN,$$
and certainly point U is on the angle bisector of $\angle MXN$.

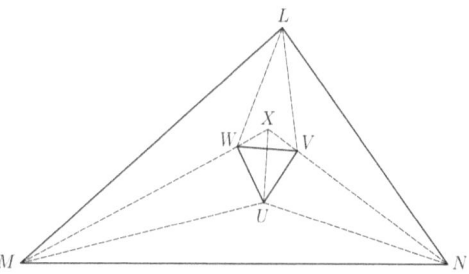

Figure MLY2

Thus point U satisfies all the conditions in the lemma below, and so U is the incenter of $\triangle XMN$. In particular,
$$\angle UMN = \angle UMW.$$
Similarly,
$$\angle WMU = \angle WML,$$
and so MW and MU are the angle trisectors of $\angle LMN$. Similarly, NU and NV are the angle trisectors of $\angle MNL$; LV and LW are the angle trisectors of $\angle NLM$.

We have shown, assuming the lemma below is valid, that the three vertices of equilateral triangle UVW are the intersections of the adjacent angle trisectors of $\triangle LMN$, which is similar to the given triangle $\triangle ABC$.

It remains to prove the lemma below.

Lemma. Let U be a point on the angle bisector XP of $\angle MXN$ in $\triangle XMN$. (Figure MLY3) Then U is the incenter of $\triangle XMN$ if and only if
$$\angle MUN = \frac{1}{2}(\angle MXN + \pi).$$

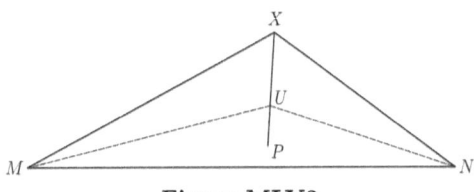

Figure MLY3

Proof. Suppose U is the incenter of $\triangle XMN$. Then because $\angle MUP$ is an exterior angle of $\triangle MUX$, we have

$$\angle MUP = \angle MXU + \angle UMX = \frac{1}{2}(\angle MXN + \angle NMX).$$

Similarly,
$$\angle NUP = \frac{1}{2}(\angle MXN + \angle MNX).$$

Therefore,
$$\begin{aligned}\angle MUN &= \angle MUP + \angle NUP \\ &= \frac{1}{2}[\angle MXN + (\angle MXN + \angle NMX + \angle MNX)] = \frac{1}{2}(\angle MXN + \pi).\end{aligned}$$

On the other hand, for point U on the angle bisector of $\angle MXN$, as U moves away from X, $\angle MUN (= \angle MUP + \angle NUP)$ increases monotonically. So, the point U that satisfies the equality

$$\angle MUN = \frac{1}{2}(\angle MXN + \pi),$$

is unique, implying that U must be the incenter of $\triangle XMN$. □

Note that the angles of $\triangle UMN$, $\triangle VNL$, $\triangle WLM$ (Figure MLY2) are

$$\{\alpha'', \beta, \gamma\}, \quad \{\beta'', \gamma, \alpha\}, \quad \{\gamma'', \alpha, \beta\},$$

respectively, where

$$\alpha'' = \alpha + \frac{2\pi}{3}, \quad \beta'' = \beta + \frac{2\pi}{3}, \quad \gamma'' = \gamma + \frac{2\pi}{3}.$$

Exercise.

(a) Show that the side length of the equilateral triangle associated with $\triangle ABC$ in the Morley theorem is

$$8R \sin \alpha \cdot \sin \beta \cdot \sin \gamma,$$

where R is the circumradius of $\triangle ABC$, and α, β, γ are as in the proof above.

The Morley Theorem

(b) Prove: The Morley theorem is true even if the (interior) angle trisectors are replaced by the exterior angle trisectors, or by the conjugate angle trisectors. (Figure MLY4) Can you imitate the proof above for these cases? And more? (See my book, *Complex Numbers and Geometry*, Mathematical Association of America, 1994.)

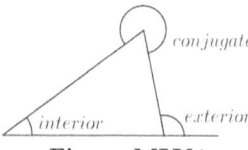

Figure MLY4

Chapter 20

Angle Trisection

20.1 Rules of Engagement

Quite often we are fascinated with geometry problems such as

1) Given a line and a point on the line, it is easy to construct the line perpendicular to the given line passing through the given point. Can you solve this problem using compass only once?

2) Given a line and a point not on the line, it is easy to construct the line parallel to the given line passing through the given point. Can you solve this problem using compass only once?

3) Given a line ℓ and a pair of points A and B on the same side of ℓ, find the point P on ℓ such that $\overline{AP} + \overline{BP}$ is minimum.

4) An equilateral triangle, a square, and a regular hexagon are easy to construct. But how about the other regular polygons?

Problems such as these are known as construction problems. To solve a construction problem, we must device an algorithm (a sequence of procedures) so that the desired figure can be obtained.

However, just like in our daily life, whether a problem can be solved or not depends very much on the tools available unless the desired figure does not exist. For example,

> Construct a convex polygon with four or more acute angles.

Exercise.[1] Why such a polygon does not exist?

In Euclidean construction, there are only two tools available, a straightedge and a compass. Furthermore, they are subject to the following constraints:

---
[1] See my book, *New Mexico Mathematics Contest Problem Book* (University of New Mexico Press, 1996), p.p. 21, 101-102.

Rule I. A straightedge can be used to construct a line segment joining two given points.

Rule II. A straightedge can be used to extend a given segment (as long as we please).

Rule III. A compass can be used to construct a circle centered at a given point passing through another given point.

Strictly speaking, the following rule should also be adopted:

Rule IV. It is possible to identify the intersection of a pair of lines, a pair of circles, or a line and a circle if they intersect.

Consequently, given line ℓ and point P not on line ℓ, to draw the line parallel to line ℓ passing through point P, we are not allowed to simply slide triangles. (Figure ATS/Left)

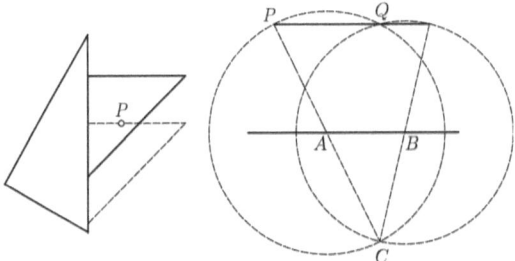

Figure ATS

Here is one way to do it. (Figure ATS/Right)

1° Choose an arbitrary pair of distinct points A and B on line ℓ, and draw circle $A(P)$ (i.e., the circle centered at A passing through point P). (Rule III)

2° Let C be the second intersection of this circle and line PA. (Rules I, II, IV)

3° Draw circle $B(C)$. (Rule III)

4° Let Q be the second intersection of these two circles. (Rule IV) Then line PQ is the desired line. (Rules I and II) (Why?)

Exercise. Solve this problem using compass only once. (There is no restriction on the number of times a straightedge is used.)

Now, Rules I, II, III, and IV imply the following.

Rule III' A compass can be used to construct a circle with the given center and the given radius.

Angle Trisection

To justify our assertion, suppose center O and radius $\overline{AB}$ are given. (Figure ATS1/Left)

1° Draw circles $O(A)$ and $A(O)$. (Rule III) Let the intersections of these two circles be C and D. (Rule IV)

2° Then clearly, CD is the perpendicular bisector of OA (Rule I), and so their intersection M (Rule IV) is the midpoint of OA.

3° Draw circle $M(B)$. (Rule III)

4° Draw segment BM (Rule I) and extend BM (Rule II) to find the (second) intersection E of line BM and circle $M(B)$. (Rule IV) Then clearly,
$$\triangle MOE \cong \triangle MAB, \quad \text{and so} \quad \overline{OE} = \overline{AB}.$$
Thus circle $O(E)$ is the desired circle.

Note carefully, our construction works even if points A, B and O are collinear (without any change).

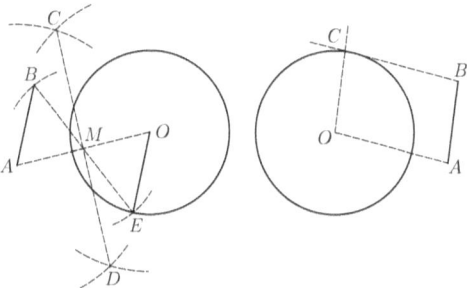

Figure ATS1

Alternatively, we can do as follows (Figure ATS1/Right): Without loss of generality, we may assume that points O, A, B are not collinear. (If these points are collinear, then draw circle $A(B)$ and choose an arbitrary point B' on the circle that is not on line OA, and replace B by B' in our discussion below.)

1° Draw the line parallel to AB passing through O. (Rules I, II, III, IV as in the example on page 164)

2° Draw the line parallel to OA passing through B. (Rules I, II, III, IV)

3° Let C be the intersection of the lines constructed in steps 1° and 2°. (Rules IV) Then $OABC$ is a parallelogram, and so $\overline{OC} = \overline{AB}$, and hence circle $O(C)$ (Rule III) is the desired circle.

Thus Rules I, II, III, IV imply Rules I, II, III', IV. Converse is trivial. Under Rules I, II, III', IV, given center O, to draw circle $O(P)$, simply choose $\overline{OP}$ as the radius.

Once we know how to draw parallel lines, given a segment of length a, then it is easy to draw segments of length na or a/n, where n is an arbitrary positive integer. (And hence also $\frac{m}{n}a$, where m and n are arbitrary positive integers.) (Figure ATS2)

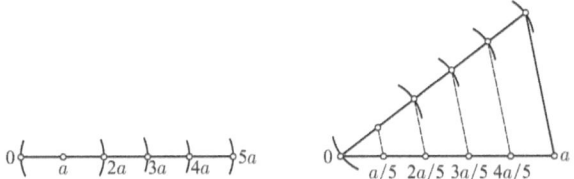

Figure ATS2

Our main purpose in this chapter is to explain why the famous angle trisection problem is impossible under Rules I, II, III', and IV, provided that straightedge and compass are allowed to be used only finitely many times. If we are allowed to use straightedge and compass infinitely many times, then clearly any angle can be trisected. For, we know an arbitrary angle can be bisected. (Figure ATS3) Furthermore, we can add or subtract angles. So the infinite series

$$\frac{1}{2} - \frac{1}{2^2} + \frac{1}{2^3} - \frac{1}{2^4} + - \cdots = \frac{1}{2} \cdot \frac{1}{1-\left(-\frac{1}{2}\right)} = \frac{1}{3}$$

tells us how to trisect a given angle (using straightedge and compass infinitely many times).

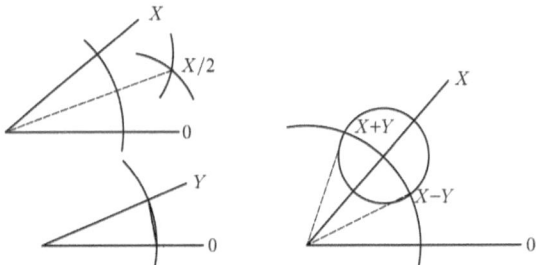

Figure ATS3

Note carefully, for any given angle, its trisectors do exist. Yet, it is impossible to construct, in general, under our constraints. By the way, there are many angles such as 90°, 45°, that can be trisected. However, this does not contradict our assertion. When we say it is impossible, we are asserting that there is no algorithm that works in every case. So, if we can exhibit one angle that cannot be trisected (using straightedge and compass finitely many times following Rules I, II, III', IV), then our proof of

Angle Trisection

impossibility is complete. (If angle θ cannot be trisected, then it is obvious that neither are $\theta/2$, $\theta/2^2$, $\theta/2^3$, $\cdots$.) Needless to say our only concern is to construct trisectors that are (theoretically) precise, not approximate ones.

20.2 The Trisection Equation

Now suppose we are given $\angle XOY$. (Figure ATS4) Draw a circle centered at O with radius r intersecting the two arms of the given angle at A and B. (As we shall see later that it is most convenient to set $r = 2$, but for the time being, we do not worry about this.) Let C be the intersection of the angle trisector (that is close to OX) and the line parallel to OX passing through B; i.e., $\angle AOC = \frac{1}{3}\angle XOY$ $(=t)$. Furthermore, let D and E be the feet of the perpendiculars from B and C to OX, respectively.

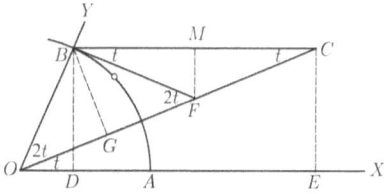

Figure ATS4

Note carefully, once radius r is fixed, then $\angle XOY$ determines OD. Converse is also true provided we restrict $\angle XOY$ to be acute. (Note that because angle $\frac{\pi}{2}$ can be trisected, and adding or subtracting a multiple of an angle that can be trisected does not affect the "trisectability" of the original angle, our restriction on the range of $\angle XOY$ does not result in a loss of generality. Besides, to prove the impossibility of the angle trisection problem, all we need is to exhibit one angle that cannot be trisected.) Same is true between $\angle XOC$ and OE. Hence we try to find the relation between OD and OE, and see whether we can solve this equation.

Let M be the midpoint of BC, and F the intersection of OC and the perpendicular bisector of BC. Then, clearly,

$$\angle CBF = \angle BCF = \angle XOC = t,$$

implying that $\angle OFB = 2t$ because $\angle OFB$ is an exterior angle of $\triangle FBC$. On the other hand, $\angle BOC = 2t$, hence $\triangle OBF$ is isosceles, and so

$$\overline{OB} = \overline{BF} = \overline{CF}\, (=r).$$

Let G be the foot of the perpendicular from B to OC. Then, clearly, $\triangle BCG$, $\triangle FCM$ and $\triangle COE$ are similar. Hence

$$\frac{\overline{CG}}{\overline{BC}} = \frac{\overline{CM}}{\overline{FC}} = \frac{\overline{OE}}{\overline{CO}}.$$

Setting
$$\overline{BM} = \overline{MC} = x, \quad \overline{OG} = \overline{GF} = y, \quad \overline{OD} = a,$$
then, $\overline{OE} = \overline{OD} + \overline{DE} = \overline{OD} + \overline{BC} = a + 2x$, and so the equalities above become
$$\frac{y+r}{2x} = \frac{x}{r} = \frac{2x+a}{2y+r}.$$

From the first equality, we obtain
$$y = \frac{2x^2}{r} - r = \frac{2x^2 - r^2}{r}.$$

Substituting this result into the second equality, we obtain
$$r(2x+a) = x(2y+r) = x\left[\frac{2(2x^2 - r^2)}{r} + r\right] = \frac{x}{r}\left(4x^2 - r^2\right).$$

Therefore,
$$4x^3 - 3r^2 x - ar^2 = 0.$$

Choosing $r = 2$, we obtain
$$x^3 - 3x - a = 0.$$

This is our trisection equation. And the trisection problem: "Given OD, can we construct OE?" becomes: Given a, can we solve this cubic equation for x with straightedge and compass (i.e., by algebraic operations that correspond to construction by straightedge and compass)?

Exercise. Can you spot the bisection equation in our derivation of the trisection equation above?

Naturally, for those who are familiar with trigonometry, because
$$\begin{aligned} \cos 3t &= \cos(2t + t) = \cos 2t \cdot \cos t - \sin 2t \cdot \sin t \\ &= \left(\cos^2 t - \sin^2 t\right) \cos t - (2 \sin t \cdot \cos t) \sin t \\ &= \cos^3 t - 3 \sin^2 t \cdot \cos t = \cos^3 t - 3\left(1 - \cos^2 t\right) \cos t \\ &= 4 \cos^3 t - 3 \cos t, \end{aligned}$$

substituting
$$\cos 3t = \frac{a}{r} = \frac{a}{2}, \quad \cos t = \frac{x}{r} = \frac{x}{2},$$
into the identity above, we obtain
$$\frac{a}{2} = 4\left(\frac{x}{2}\right)^3 - 3\left(\frac{x}{2}\right), \quad x^3 - 3x - a = 0;$$
which is our trisection equation.

Angle Trisection

20.3 Computations by Straightedge and Compass

Our next step is to investigate what algebraic operations can be performed by straightedge and compass.

(i) Addition and Subtraction.
Given segments of lengths a and b, we can construct segments of lengths $a + b$ and $a - b$. (Figure ATS5)

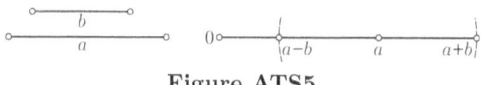

Figure ATS5

(ii) Multiplication and Division
We need the following known as the power theorem.

> *Theorem 1.* Let P be the intersection of chords AB and CD (or their extensions) of a circle. (Figure ATS6/Left and Middle) Then
> $$\overline{PA} \cdot \overline{PB} = \overline{PC} \cdot \overline{PD}.$$

Proof. Because points A, B, C, D are cocyclic, we have $\angle PAD = \angle PCB$. Furthermore, $\angle APD = \angle CPB$. It follows that $\triangle PAD \sim \triangle PCB$. Therefore,
$$\frac{\overline{PA}}{\overline{PD}} = \frac{\overline{PC}}{\overline{PB}}, \quad \overline{PA} \cdot \overline{PB} = \overline{PC} \cdot \overline{PD}.$$
□

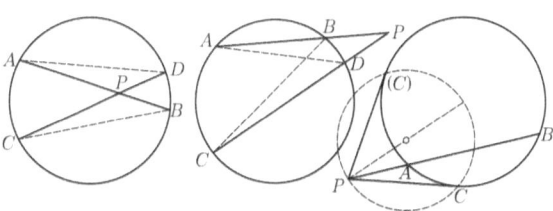

Figure ATS6

Corollary 2. Let P be a point that is outside a circle, and a line through P intersects the circle at points A and B. Furthermore, let C be a point on the circle such that PC is tangent to the circle. (Figure ATS6/Right) Then
$$\overline{PA} \cdot \overline{PB} = \overline{PC}^2.$$

Proof. The corollary follows immediately from the theorem by simply rotating the line PCD (in the theorem) around point P so that points C and D coincide. □

Exercises.

1. Conversely, given four points A, B, C, D such that
$$\overline{PA} \cdot \overline{PB} = \overline{PC} \cdot \overline{PD},$$
where P is the intersection of AB and CD, is it necessary that points A, B, C, D be cocyclic?

2. Given a pair of lines and a point not on the given lines, construct a circle tangent to the pair of given lines and passing through the given point. Is such a circle unique?

3. Given a line and a pair of points on the same side of the given line, construct a circle tangent to the given line and passing through the pair of given points. Is such a circle unique?

With the help of the power theorem, we can construct the product and the quotient of two given segments of lengths a and b. (Figure ATS7/Left)

1° Choose three collinear points P, A, B such that $\overline{PA} = a$, $\overline{PB} = b$.

2° Draw an arbitrary line passing through P (but not A nor B), and let D be a point on this line such that $\overline{PD} = 1$.

3° Draw the circle passing through points A, B and D. (This is possible because these three points are not collinear.) (What if $a = b$; i.e., A and B coincide?)

4° Let C be the second intersection of the circle constructed in step 3° and line PD. Then, by the power theorem,
$$\overline{PC} \cdot \overline{PD} = \overline{PA} \cdot \overline{PB}; \text{ i.e., } \overline{PC} = ab.$$

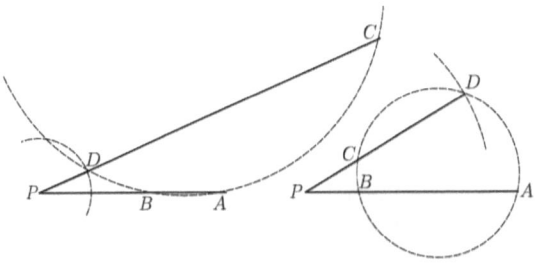

Figure ATS7

Angle Trisection

Construction for the division is similar, except this time we must assume that $b \neq 0$. (Figure ATS7; Right) Then simply choose

$$\overline{PA} = a, \quad \overline{PB} = 1, \quad \overline{PD} = b.$$

And the rest is the same as the case for the product. Note that if $b = 0$, then points A, B, D are collinear, and so there exists no circle that passes through these three points.

(iii) Square Root.

Given a segment of length a, we can construct a segment of length $\sqrt{a}$.

1° Let P, A, B be collinear points (where A and B are on the same side of point P) such that $\overline{PA} = a$, $\overline{PB} = 1$. (Figure ATS8/Left)

2° Construct an arbitrary circle passing through A and B. (Here, we are assuming that points A and B are distinct; i.e., $a \neq 1$. Otherwise, the problem is trivial.)

3° Finally, from point P, draw a tangent to the circle constructed in step 2°. Then, by the corollary of the power theorem, we have

$$\overline{PT}^2 = \overline{PA} \cdot \overline{PB}, \quad \overline{PT} = \sqrt{a},$$

where T is the point that the tangent line from P touches the circle.

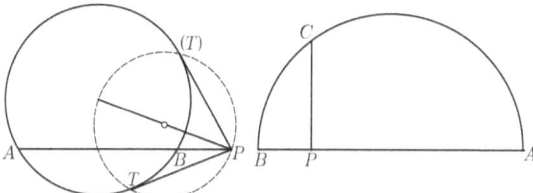

Figure ATS8

Remark. In step 1° above, we assumed that points A and B are on the same side of point P. We could also let point P be between points A and B (and $\overline{PA} = a$, $\overline{PB} = 1$ as before). (Figure ATS8/Right) Then draw the circle with diameter AB, and let C be an intersection of the circle and the perpendicular to diameter AB passing through P. Again, by the power theorem, we have

$$\overline{PC}^2 = \overline{PA} \cdot \overline{PB}, \quad \overline{PC} = \sqrt{a}.$$

Summing up, we can construct any expression that involves the rational operations (i.e., addition, subtraction, multiplication and division), and square root finitely many times by using straightedge and compass finitely many times.

20.4 Fields and their Extensions

Now a line segment is determined by the two endpoints, and a line can be described by two points on it. A circle can be described by its center and a point on it, or three points on it. A circular arc can be described by the two endpoints and the center or another point on it. Thus in a construction problem, a given figure F (assuming a figure is given) can be described by a finite number of points. (For example, an angle can be described by three points.) Furthermore, two of these points can be designated as the origin $(0, 0)$ and the unit point $(1, 0)$ of a coordinate plane (provided the given figure F contains at least two points). Because each point can be described by an ordered pair of real numbers, the given figure F can be described by a finite number of real numbers.

As we have seen above, all the rational operations (i.e., addition, subtraction, multiplication and division, except division by 0) can be carried out by straightedge and compass, we can construct all the rational points (i.e., points with rational coordinates).

A set K in which the rational operations can be carried out freely (except division by 0, as always) is called a *field*. More precisely, a set K is called a field if the following conditions are satisfied:

(i) For any pair of elements a and b in K, their sum, denoted by $a+b$, is defined and belongs to K.

(ii) Commutative law of addition:
$$a + b = b + a \quad \text{for all } a \text{ and } b \text{ in } K.$$

(iii) Associative law of addition:
$$(a + b) + c = a + (b + c) \quad \text{for all } a, b \text{ and } c \text{ in } K.$$

(iv) There exists an element 0 in K (called the *additive identity* or *zero*) such that
$$a + 0 = a \quad \text{for all } a \text{ in } K.$$

(v) For every element a in K, there exists an element in K (called the *additive inverse* of a), denoted by $(-a)$, such that
$$a + (-a) = 0.$$

(i') For any pair of elements a and b in K, their product, denoted by $a \cdot b$, is defined and belongs to K.

(ii') Commutative law of multiplication:
$$a \cdot b = b \cdot a \quad \text{for all } a \text{ and } b \text{ in } K.$$

Angle Trisection

(iii') Associative law of multiplication:
$$(a \cdot b) \cdot c = a \cdot (b \cdot c) \quad \text{for all } a, b \text{ and } c \text{ in } K.$$

(iv') There exists an element 1 in K (called the *multiplicative identity* or *unit*), that is not equal to 0, such that
$$a \cdot 1 = a \quad \text{for all } a \text{ in } K.$$

(v') For every element a in K, except $a = 0$, there exists an element in K (called the *multiplicative inverse* of a), denoted by a^{-1}, such that
$$a \cdot a^{-1} = 1.$$

(vi) Distributive law:
$$a \cdot (b + c) = a \cdot b + a \cdot c \quad \text{for every } a, b, c \text{ in } K.$$

Remark. It is easy to prove that the additive identity 0, the multiplicative identity 1, and for every element a in K, its additive inverse $(-a)$ and multiplicative inverse a^{-1} (provided $a \neq 0$) are all unique. For example, suppose there exist two additive identities, 0 and $0'$, then
$$0 = 0 + 0' = 0' + 0 = 0'.$$

Suppose in a description of the given figure F, a non-rational number t is involved, then we can extend the rational field to include t and still preserve the structure of a field. All we have to do is to consider the set of all elements of the form
$$\frac{a_0 t^m + a_1 t^{m-1} + a_2 t^{m-2} + \cdots + a_{m-1} t + a_m}{b_0 t^n + b_1 t^{n-1} + b_2 t^{n-2} + \cdots + b_{n-1} t + b_n},$$
where the coefficients $a_0, a_1, \cdots, a_m, b_0, b_1, \cdots, b_n$ are rational numbers, m and n are some nonnegative integers, and we assume that the denominator is not zero. It is obvious that the set of all such "rational expressions" forms a field. Furthermore, this new extended field contains the rational field as its "subfield". We denote this extended field as $K(t)$. Note carefully, all the elements of $K(t)$ can be constructed by straightedge and compass starting from the given figure F (because t is given in the figure F).

Suppose there still exists a number, say t_1, in the description of the given figure F, that does not belong to the extended field $K(t)$. Then we can further extend $K(t)$ by considering all the rational expressions of the form
$$\frac{a'_0 t_1^i + a'_1 t_1^{i-1} + a'_2 t_1^{i-2} + \cdots + a'_{i-1} t_1 + a'_i}{b'_0 t_1^j + b'_1 t_1^{j-1} + b'_2 t_1^{j-2} + \cdots + b'_{j-1} t_1 + b'_j},$$
where all the coefficients are elements of $K(t)$, i and j are some nonnegative integers, and the denominator is not zero. We denote the field of all such expressions as

$K(t, t_1)$. Again all the elements of $K(t, t_1)$ can be constructed by straightedge and compass starting from the given figure F. Note carefully, $K(t, t_1) = K(t_1, t)$; i.e., the order that the numbers t and t_1 are added is irrelevant. (Why?)

Repeating this process finitely many times (if necessary), we obtain, eventually, the smallest field, denoted by K_0, that includes all the real numbers involved in the description of the given figure F (because F can be described by a finite number of real numbers). Note that all the elements of K_0 can be constructed by straightedge and compass starting from the given figure F.

Next, observe that the line passing through a pair of points $A(a_1, a_2)$ and $B(b_1, b_2)$ in the given figure F can be expressed by the equation

$$ax + by + c = 0,$$

where coefficients $a = a_2 - b_2$, $b = -a_1 + b_1$, $c = a_1b_2 - a_2b_1$, are elements of field K_0. And to find the intersection of two lines

$$ax + by + c = 0 \quad \text{and} \quad a'x + b'y + c' = 0,$$

is tantamount to solving this system of simultaneous equations. We obtain

$$x = -\frac{cb' - c'b}{ab' - a'b}, \quad y = -\frac{ac' - a'c}{ab' - a'b},$$

unless $ab' - a'b = 0$ (i.e., two lines are parallel), in which case either there is no intersection or two lines coincide (hence every point on the line is an intersection). Note that these expressions are elements of field K_0, implying that using straightedge only, we cannot obtain a real number that is not in field K_0.

Next a circle in a given figure F can be expressed by equation of the form

$$x^2 + y^2 + 2\alpha x + 2\beta y + \gamma = 0,$$

where coefficients 2α, 2β, γ are elements of K_0. (Why?) To find the intersection of this circle and line

$$ax + by + c = 0 \quad \text{(where } a \text{ and } b \text{ can not be both zero)},$$

is tantamount to solving this system of simultaneous equations. Assuming $b \neq 0$ (the case $b = 0$ is easier, and is left as an exercise for the reader), and substituting the expression

$$y = -\frac{ax + c}{b}$$

into the equation of the circle, we obtain

$$Ax^2 + Bx + C = 0,$$

where coefficients

$$A = a^2 + b^2 \neq 0, \quad B = 2(ac + \alpha b^2 - \beta ab), \quad C = c^2 - 2\beta bc + \gamma b^2$$

Angle Trisection

are all elements of field K_0. Solving this quadratic equation in x, we obtain

$$x = \frac{-B \pm \sqrt{B^2 - 4AC}}{2A}.$$

Naturally, if $B^2 - 4AC < 0$, then the circle and the line do not intersect. Now suppose $B^2 - 4AC \geq 0$. Then $(-B \pm \sqrt{B^2 - 4AC})/(2A)$, or rather, $\sqrt{B^2 - 4AC}$, may or may not be in K_0. However, even if $\sqrt{B^2 - 4AC}$ is not in K_0, this quantity can be constructed using straightedge and compass as we saw above. A similar situation happens for y.

As for the intersection of two circles

$$x^2 + y^2 + 2\alpha x + 2\beta y + \gamma = 0 \quad \text{and} \quad x^2 + y^2 + 2\alpha' x + 2\beta' y + \gamma' = 0;$$

taking the difference of these two equations, we obtain

$$2\alpha_0 x + 2\beta_0 y + \gamma_0 = 0,$$

where coefficients $\alpha_0 = \alpha - \alpha'$, $\beta_0 = \beta - \beta'$, $\gamma_0 = \gamma - \gamma'$, are all elements in K_0. This is an equation of the line passing through two intersections of the circles provided the circles intersect. (Even if the circles do not intersect, this equation represents a line that has a geometric meaning; but that is not our concern now.) So the situation is exactly the same as the case of a circle and a line.

In either case, the coordinates of the intersections of two circles or a circle and a line are either elements of K_0 or of the form

$$p + q\sqrt{k} \quad \text{(where } p, q, k \text{ are in } K_0, \text{ but } \sqrt{k} \text{ is not)}.$$

Luckily, numbers of this form constitute a field. For,

$$\left(p + q\sqrt{k}\right) \pm \left(p' + q'\sqrt{k}\right) = (p \pm p') + (q \pm q')\sqrt{k},$$
$$\left(p + q\sqrt{k}\right) \cdot \left(p' + q'\sqrt{k}\right) = (pp' + qq'k) + (pq' + p'q)\sqrt{k}.$$

Note that $p \pm p'$, $q \pm q'$, $pp' + qq'k$, $pq' + p'q$, are all elements of K_0. And for division, we may assume that $q' \neq 0$. (The case $q' = 0$ is trivial.) Then

$$\frac{p + q\sqrt{k}}{p' + q'\sqrt{k}} = \frac{p + q\sqrt{k}}{p' + q'\sqrt{k}} \cdot \frac{p' - q'\sqrt{k}}{p' - q'\sqrt{k}} = \frac{(pp' - qq'k) + (p'q - pq')\sqrt{k}}{p'^2 - q'^2 k}.$$

Note that $p'^2 - q'^2 k \neq 0$; because otherwise $\sqrt{k} = \pm p'/q'$ would be an element of K_0, contradicting our assumption. Verification of the rest of the conditions is straightforward. Clearly, this new field contains field K_0 as its subfield. We shall denote this extended field as K_1.

It may happen that we need to draw a line or a circle using one or more of the new intersections. In which case the coefficients of the line or the circle are all elements

of K_1. While the coordinates of the intersection of two such lines are elements of K_1, and that of the intersections of two such circles, or a circle and a line could be numbers of the form
$$p + q\sqrt{k_1} \quad \text{(where } p, q, k_1 \text{ are in } K_1, \text{ but } \sqrt{k_1} \text{ is not)}.$$
In such a case, repeating the same argument as before, we see that the set of all such numbers constitutes a field containing field K_1 as its subfield; we denote this newly extended field by K_2.

Because straightedge and compass are allowed to be used only finitely many times, repeating this process if necessary, we see that the coordinates of all constructible points must be elements of K_n for some integer n.

20.5 Impossibility Proofs

We can now complete the proof of the impossibility of trisecting an arbitrary angle using straightedge and compass finitely many times. Recall that our trisection equation is
$$x^3 - 3x - a = 0,$$
where we used radius $r = 2$. (Figure ATS4) We choose $a = 1$; i.e., we want to show that 60°-angle cannot be trisected. Assume to the contrary that it can be trisected. Then a root of
$$x^3 - 3x - 1 = 0$$
must be an element of K_n for some n. Let n be the minimum integer such that K_n contains a root of the trisection equation; i.e.,
$$x = p + q\sqrt{k} \quad \text{(where } p, q, k \text{ are in } K_{n-1}, \text{ but } \sqrt{k} \text{ is not)}.$$
Substituting this expression into our trisection equation, we obtain
$$\left(p + q\sqrt{k}\right)^3 - 3\left(p + q\sqrt{k}\right) - 1 = 0.$$
Rewriting, we obtain
$$u + v\sqrt{k} = 0,$$
where
$$u = p^3 + 3pq^2k - 3p - 1 \quad \text{and} \quad v = 3p^2q + q^3k - 3q$$
are elements of K_{n-1}. Because $\sqrt{k}$ is not an element of K_{n-1}, these two expressions must both be zero. In other words,
$$u + v\sqrt{k} = 0 \iff u = v = 0.$$
This is easy to prove. Suppose $v \neq 0$. Then
$$\sqrt{k} = -\frac{u}{v},$$

Angle Trisection

where the right-hand side is an element of K_{n-1}, but the left-hand side is not. The contradiction shows that our assumption that $v \neq 0$ is not tenable. Thus $v = 0$, which, in turn, implies $u = 0$. So we must have

$$u = p^3 + 3pq^2k - 3p - 1 = 0 \quad \text{and} \quad v = 3p^2q + q^3k - 3q = 0.$$

But then $x = p - q\sqrt{k}$ must also be a root of our trisection equation. For, substituting this expression into the left-hand side of the trisection equation, we obtain

$$\left(p - q\sqrt{k}\right)^3 - 3\left(p - q\sqrt{k}\right) - 1$$
$$= (p^3 + 3pq^2k - 3p - 1) - (3p^2q + q^3k - 3q)\sqrt{k}$$
$$= u - v\sqrt{k} = 0.$$

Let w be the third root of the trisection equation; i.e.,

$$x^3 - 3x - 1 = \left[x - \left(p + q\sqrt{k}\right)\right] \cdot \left[x - \left(p - q\sqrt{k}\right)\right] \cdot (x - w).$$

Comparing the coefficients of x^2 on both sides, we obtain

$$\left(p + q\sqrt{k}\right) + \left(p - q\sqrt{k}\right) + w = 0; \quad \text{i.e.,} \quad w = -2p,$$

which shows that the trisection equation has a root w that is in K_{n-1}, contradicting our assumption that n is the minimum integer such that K_n contains a root of the trisection equation. The contradiction shows that 60°-angle cannot be trisected.

Example 1. We know that 90°-angle can be trisected because a 60°-angle can be constructed. And, in fact, in this case $a = r\cos(\angle XOY) = 2\cos 90° = 0$ and the trisection equation becomes

$$x^3 - 3x = 0; \quad \text{i.e.,} \quad x\left(x - \sqrt{3}\right)\left(x + \sqrt{3}\right) = 0.$$

Now, for $\angle XOY = 90°$, we have

$$x = r\cos\left(\frac{1}{3}\angle XOY\right) = 2\cos 30° = \sqrt{3},$$

which is obviously constructible with straightedge and compass. That explains the root $\sqrt{3}$. But what about the other two roots, 0 and $-\sqrt{3}$? Note that $\angle XOY = 90°$ is not the only angle that gives $a = 2\cos(\angle XOY) = 0$. If $\angle XOY = 270°$, then $\frac{1}{3}\angle XOY = 90°$, and so

$$a = 2\cos(\angle XOY) = 2\cos 270° = 0 \quad \text{and} \quad x = 2\cos\left(\frac{1}{3}\angle XOY\right) = 2\cos 90° = 0.$$

The remaining root $-\sqrt{3}$ comes from $\angle XOY = 450°$; because

$$2\cos(\angle XOY) = 2\cos 450° = 0, \quad \text{and} \quad 2\cos\left(\frac{1}{3}\angle XOY\right) = 2\cos 150° = -\sqrt{3}.$$

To be precise, all the angles of the form

$$\angle XOY = \pm(90° + n \cdot 360°),$$

where n is an integer, share the same trisection equation, and

$$x = \sqrt{3}, \ -\sqrt{3}, \ \text{or} \ 0,$$

depending on whether n gives remainder 0, 1, or 2, when divided by 3.

Example 2. Clearly, 45°- and 30°-angles are constructible, so is $15° = 45° - 30°$. Hence $3 \times 15° = 45°$ can be trisected. In this case, $2\cos 45° = \sqrt{2}$, and the trisection equation becomes

$$x^3 - 3x - \sqrt{2} = 0,$$

which can be factored as

$$\left(x + \sqrt{2}\right)\left(x - \frac{1+\sqrt{3}}{\sqrt{2}}\right)\left(x - \frac{1-\sqrt{3}}{\sqrt{2}}\right) = 0.$$

Clearly, all the roots are constructible.

Exercise.

(a) For what $\angle XOY$, do we have

$$x = 2\cos\left(\frac{1}{3}\angle XOY\right) = -\sqrt{2}?$$

(b) What about the other two roots?

Exactly the same trick works for another classical Greek problem, known as the problem of Delos; i.e., to double the volume of a given cube. We may assume that an edge of the given cube is of unit length (hence its volume is the unit volume), and we are required to construct the edge of length x that satisfies the equation

$$x^3 - 2 = 0.$$

Suppose a segment of length x satisfying this equation can be constructed using straightedge and compass finitely many times. Then x must be an element of K_n for some n. Again we assume that n is the minimum integer such that K_n contains a root of this "doubling equation". Then x can be expressed as

$$x = p + q\sqrt{k} \quad \text{(where } p, q, k \text{ are elements of } K_{n-1}, \text{ but } \sqrt{k} \text{ is not)}.$$

Angle Trisection

Substituting this expression into the doubling equation, we obtain

$$\left(p + q\sqrt{k}\right)^3 - 2 = 0; \text{ i.e., } u + v\sqrt{k} = 0,$$

where

$$u = p^3 + 3pq^2k - 2, \quad v = 3p^2q + q^3k$$

are in K_{n-1}. As before, we must have $u = v = 0$ (because $\sqrt{k}$ is not an element of K_{n-1}). But then $x = p - q\sqrt{k}$ must also be a root of $x^3 - 2 = 0$, because

$$\left(p - q\sqrt{k}\right)^3 - 2 = u - v\sqrt{k} = 0.$$

It follows that, for some w,

$$x^3 - 2 = \left[x - \left(p + q\sqrt{k}\right)\right] \cdot \left[x - \left(p - q\sqrt{k}\right)\right] \cdot (x - w).$$

Comparing the coefficients of x^2 on both sides, we obtain

$$\left(p + q\sqrt{k}\right) + \left(p - q\sqrt{k}\right) + w = 0; \text{ i.e., } w = -2p,$$

implying that the root $w = -2p$ is an element of K_{n-1}. This contradicts our assumption that n is the minimum such integer. The contradiction shows that the length of the edge satisfying the doubling equation is not constructible.

Having used the same trick twice, it is time to formulate a general theorem. Because its proof is essentially the repetition of the trick above, we leave it as an exercise for the reader.

Theorem 3. Suppose a cubic polynomial with rational coefficients has no rational root, then none of its roots is constructible starting from the rational field.

Naturally, we could have used the trick to prove this theorem first, then apply the theorem to the trisection equation (for 60°-angle) and the doubling equation. Because these two equations are both monic (i.e., the leading coefficient is 1), and all coefficients are integers, a rational root, if any, must be an integer. (Why?) So all we need is to show that an integer root does not exist. In the case of the trisection equation (for 60°-angle), setting

$$f(x) = x^3 - 3x - 1,$$

we have

$$f(-2) = -3 < 0, \ f(-1) = 1 > 0, \ f(0) = -1 < 0, \ f(1) = -3 < 0, \ f(2) = 1 > 0.$$

And because $f(x)$ is a continuous function, by the intermediate value theorem, $f(x) = 0$ has exactly one root in each of the intervals $(-2, -1)$, $(-1, 0)$, $(1, 2)$, implying that it has no integer root (hence neither rational root).

180   Mathemagical Buffet

In the case of the doubling equation, setting $g(x) = x^3 - 2$, we see immediately that $g(x)$ is a monotone increasing continuous function even without appealing to calculus (because the first term is a monotone increasing continuous function, and the second term is a constant). Furthermore,

$$g(1) = -1 < 0, \quad g(2) = 6 > 0,$$

so the unique real root is in the interval (1, 2). Therefore, $g(x) = x^3 - 2 = 0$ has no integer root (nor rational root).

Exercise. Give alternative proofs (i.e., without using continuity) that neither the trisection equation (for 60°-angle) nor the doubling equation has a rational root.

As for the remaining classical Greek problem; i.e., to construct a square whose area is equal to the area of a given circle, we need to show that π is transcendental (proved by C.L.F. Lindemann in 1882). Unfortunately, that is beyond our scope; hence will not be discussed here.

20.6   Bending of the Rules

So far the use of the straightedge is subject to Rules I and II (page 164). Now if we are allowed to mark two points, say P and Q, on the straightedge (like a ruler), then it is possible to trisect an angle as shown by Archimedes.

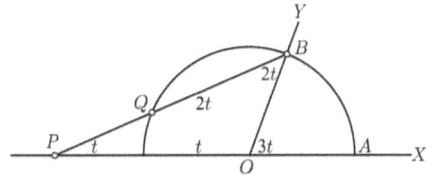

Figure ATS9

Given an angle XOY, draw a circle centered at O and radius $\overline{PQ}$ intersecting the two legs of $\angle XOY$ at A and B. (Figure ATS9) Keeping point P on the extension of AO, slide the marked straightedge so that Q is on the circle and the straightedge passes through point B. Then we have $\angle APB = \frac{1}{3}\angle AOB$. Proof is essentially the same with our derivation of the trisection equation, and hence left for the reader.

Exercises.

1. Construct a 30-degree angle by folding a given rectangular paper.

2. Given $\angle XOY < \angle XOZ = 90°$ (Figure ATS10), choose two points A, B on OZ such that A is the midpoint of OB. Draw line $AC \parallel OX$. Then fold the

paper (along the dotted line) such that point O falls on line AC, and point B falls on line OY. Show that

$$\angle XOD = \frac{1}{3} \angle XOY.$$

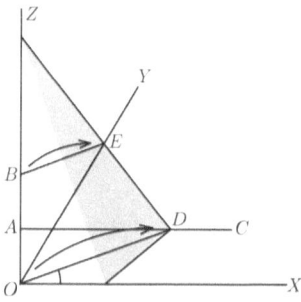

Figure ATS10

3. Given two points A and B, find the locus of point P such that

$$2\angle PAB = \angle PBA.$$

Can you use this curve to trisect a given angle?

As for the problem of doubling the volume, an ancient Greek mathematician Menächmus (375 - 325 B.C.) solved it using parabolas. In modern terms, the y-coordinate of the intersection (other than the origin) of $2y = x^2$ and $x = y^2$ gives the desired length $y = \sqrt[3]{2}$. ($x = \sqrt[3]{4}$)

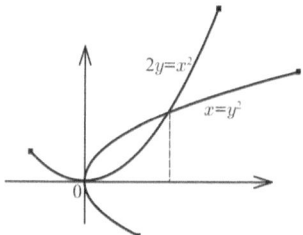

Figure ATS11

20.7 Regular Polygons

Constructions of an equilateral triangle, square and regular hexagon are well-known. But how about other regular polygons, say a regular pentagon? It turns out that HI-

RANO Yoshifusa of Japan discovered a very simple construction of a regular pentagon in 1864.

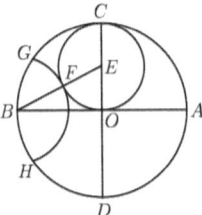

Figure ATS12

Let AB and CD be a pair of mutually perpendicular diameters of the unit circle O. (Figure ATS12) Draw the circle with diameter OC, and connect its center E with B. Let F be the intersection of segment BE and circle E. Now draw circle $B(F)$ (i.e., the circle centered at B passing through F). Let G and H be the intersections of circle $B(F)$ and the unit circle O. Then GH is the side of the regular pentagon inscribed in the unit circle O opposite the vertex A. Proof is straightforward. Because

$$\overline{BE} = \sqrt{\overline{BO}^2 + \overline{OE}^2} = \sqrt{1 + \left(\frac{1}{2}\right)^2} = \frac{\sqrt{5}}{2}.$$

$$\overline{BG} = \overline{BF} = \overline{BE} - \overline{EF} = \frac{\sqrt{5}-1}{2}.$$

But $\angle AGB = 90°$, and so

$$\cos(\angle ABG) = \frac{\overline{BG}}{\overline{AB}} = \frac{\sqrt{5}-1}{4}.$$

Therefore,

$$\angle ABG = \frac{2\pi}{5} = 72°. \quad (\text{See } Remark \text{ below.})$$

Because $\triangle OBG$ is isosceles, we have

$$\angle BOG = \pi - (\angle OBG + \angle OGB) = \pi - \frac{2\pi}{5} \times 2 = \frac{\pi}{5}, \quad \angle GOH = \frac{2\pi}{5}.$$

Thus GH is precisely the length of the regular pentagon inscribed in the unit circle O.

Exercise. Knowing that an equilateral triangle and a regular pentagon are constructible, how do we construct a regular 15-gon? (*Hint:* $\frac{1}{15} = \frac{2}{5} - \frac{1}{3}$ or $\frac{1}{15} = \frac{2}{3} - \frac{3}{5}$.)

Angle Trisection

Remark. Suppose a regular pentagon is inscribed in the unit circle (in the complex plane). Then the five vertices divide the circle into five equal parts. (Figure ATS13) Assuming one of the vertices is at $z = 1$, they satisfy the equation[2]

$$z^5 - 1 = (z-1)\left(z^4 + z^3 + z^2 + z + 1\right) = 0.$$

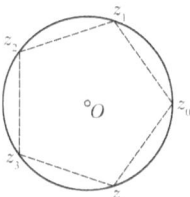

Figure ATS13

Excluding the vertex $z = 1$, the remaining four vertices satisfy the equation

$$z^4 + z^3 + z^2 + z + 1 = 0.$$

Dividing both sides by z^2, we obtain

$$\left(z^2 + \frac{1}{z^2}\right) + \left(z + \frac{1}{z}\right) + 1 = 0.$$

Substituting

$$z^2 + \frac{1}{z^2} = \left(z + \frac{1}{z}\right)^2 - 2,$$

we obtain

$$\left(z + \frac{1}{z}\right)^2 + \left(z + \frac{1}{z}\right) - 1 = 0.$$

Now

$$z_1 = \cos\frac{2\pi}{5} + i\sin\frac{2\pi}{5}$$

is one of the four roots, and

$$z_1 + \frac{1}{z_1} = \left(\cos\frac{2\pi}{5} + i\sin\frac{2\pi}{5}\right) + \left(\cos\frac{2\pi}{5} - i\sin\frac{2\pi}{5}\right) = 2\cos\frac{2\pi}{5}.$$

And so we obtain

$$\left(2\cos\frac{2\pi}{5}\right)^2 + \left(2\cos\frac{2\pi}{5}\right) - 1 = 0.$$

---

[2] See my book, *Complex Numbers and Geometry* (Mathematical Association of America, 1994), p. 37.

This is (only) a quadratic equation with integer coefficients, so its roots must be constructible; in fact,

$$2\cos\frac{2\pi}{5} = \frac{-1 \pm \sqrt{1^2+4}}{2} = \frac{-1 \pm \sqrt{5}}{2}.$$

But $\cos\frac{2\pi}{5} > 0$, and so

$$\cos\frac{2\pi}{5} = \frac{\sqrt{5}-1}{4},$$

which confirms again that a regular pentagon is constructible.

Now, we can't resist the temptation of applying this same method to regular heptagon. This time, we obtain

$$z^7 - 1 = (z-1)\left(z^6 + z^5 + z^4 + z^3 + z^2 + z + 1\right) = 0.$$

Dividing both sides of

$$z^6 + z^5 + z^4 + z^3 + z^2 + z + 1 = 0$$

by z^3, we obtain

$$\left(z^3 + \frac{1}{z^3}\right) + \left(z^2 + \frac{1}{z^2}\right) + \left(z + \frac{1}{z}\right) + 1 = 0,$$

$$\left(z + \frac{1}{z}\right)^3 + \left(z + \frac{1}{z}\right)^2 - 2\left(z + \frac{1}{z}\right) - 1 = 0.$$

Setting $z = \cos\frac{2\pi}{7} + i\sin\frac{2\pi}{7}$, we obtain

$$z + \frac{1}{z} = \left(\cos\frac{2\pi}{7} + i\sin\frac{2\pi}{7}\right) + \left(\cos\frac{2\pi}{7} - i\sin\frac{2\pi}{7}\right) = 2\cos\frac{2\pi}{7}.$$

Therefore,

$$\left(2\cos\frac{2\pi}{7}\right)^3 + \left(2\cos\frac{2\pi}{7}\right)^2 - 2\left(2\cos\frac{2\pi}{7}\right) - 1 = 0.$$

This is a cubic polynomial in $x = 2\cos\frac{2\pi}{7}$ with integer coefficients, so if a regular heptagon is constructible, then

$$f(x) = x^3 + x^2 - 2x - 1$$

must have a rational root (by Theorem 3, page 179). Suppose $x = \frac{u}{v}$ is a root (where u and v are relatively prime integers). Then

$$\left(\frac{u}{v}\right)^3 + \left(\frac{u}{v}\right)^2 - 2\left(\frac{u}{v}\right) - 1 = 0.$$

Multiplying both sides by v^3, and rewriting, we obtain

$$u^3 = -v\left(u^2 - 2uv - v^2\right) \quad \text{and} \quad v^3 = u\left(u^2 + uv - 2v^2\right).$$

Angle Trisection

The first equality shows that v divides u (but u and v are relatively prime), and so we must have $v = \pm 1$. Similarly, the second equality shows that u divides v, and so we must have $u = \pm 1$. Hence the only possible rational roots are $x = \pm 1$. But

$$f(1) = 1^3 + 1^2 - 2 - 1 = -1 \quad \text{and} \quad f(-1) = (-1)^3 + (-1)^2 - 2(-1) - 1 = 1,$$

and so $f(x) = 0$ has no rational root, implying that a regular heptagon is not constructible.

20.8   Regular 17-gon

In 1796, C.F. Gauss found that a regular 17-gon can be constructed. The 19-year old Gauss was so delighted with his discovery that, according to the legend, he decided to be a mathematician.

Let $\omega = \cos\theta + i\sin\theta$, where $\theta = \frac{2\pi}{17}$. Then

$$\omega^{17} = 1 \quad \text{and} \quad 1 + \omega + \omega^2 + \omega^3 + \cdots + \omega^{15} + \omega^{16} = 0.$$

Set

$$C(k) = \frac{1}{2}\left(\omega^k + \frac{1}{\omega^k}\right) = \cos k\theta.$$

Then it is simple to verify that

$$\begin{aligned} C(-k) &= C(k) = C(17-k), \\ 2C(m)C(n) &= C(m-n) + C(m+n), \\ C(1) + C(2) &+ C(3) + \cdots + C(7) + C(8) = -\frac{1}{2}. \end{aligned}$$

Let

$$\begin{aligned} a &= C(1) + C(4), \quad b = C(3) + C(5), \\ c &= C(2) + C(8), \quad d = C(6) + C(7); \\ e &= a + c, \quad f = b + d. \quad \text{Then} \quad e + f = -\frac{1}{2}. \end{aligned}$$

Furthermore,

$$\begin{aligned} 2ab &= 2\{C(1) + C(4)\} \cdot \{C(3) + C(5)\} \\ &= 2\{C(1)C(3) + C(4)C(3) + C(1)C(5) + C(4)C(5)\} \\ &= \{C(2) + C(4)\} + \{C(1) + C(7)\} + \{C(4) + C(6)\} + \{C(1) + C(9)\} \\ &= \{C(2) + C(9)\} + 2\{C(4) + C(1)\} + \{C(7) + C(6)\} \\ &= c + 2a + d = 2a + c + d. \\ 2ac &= 2\{C(1) + C(4)\} \cdot \{C(2) + C(8)\} \end{aligned}$$

$$
\begin{aligned}
&= 2\{C(1)C(2) + C(4)C(2) + C(1)C(8) + C(4)C(8)\} \\
&= \{C(1) + C(3)\} + \{C(2) + C(6)\} + \{C(7) + C(9)\} + \{C(4) + C(12)\} \\
&= C(1) + C(2) + C(3) + C(4) + C(12) + C(6) + C(7) + C(9) \\
&= C(1) + C(2) + C(3) + C(4) + C(5) + C(6) + C(7) + C(8) = -\frac{1}{2}. \\
2ad &= 2\{C(1) + C(4)\} \cdot \{C(6) + C(7)\} \\
&= 2\{C(1)C(6) + C(1)C(7) + C(4)C(6) + C(4)C(7)\} \\
&= \{C(5) + C(7)\} + \{C(6) + C(8)\} + \{C(2) + C(10)\} + \{C(3) + C(11)\} \\
&= \{C(5) + C(3)\} + \{C(7) + C(6)\} + \{C(8) + C(2)\} + \{C(10) + C(11)\} \\
&= b + d + c + d = b + c + 2d. \\
2bc &= 2\{C(3) + C(5)\} \cdot \{C(2) + C(8)\} \\
&= 2\{C(2)C(3) + C(2)C(5) + C(8)C(3) + C(8)C(5)\} \\
&= \{C(1) + C(5)\} + \{C(3) + C(7)\} + \{C(5) + C(11)\} + \{C(3) + C(13)\} \\
&= \{C(1) + C(13)\} + 2\{C(5) + C(3)\} + \{C(7) + C(11)\} \\
&= a + 2b + d. \\
2bd &= 2\{C(3) + C(5)\} \cdot \{C(6) + C(7)\} \\
&= 2\{C(3)C(6) + C(5)C(6) + C(3)C(7) + C(5)C(7)\} \\
&= \{C(3) + C(9)\} + \{C(1) + C(11)\} + \{C(4) + C(10)\} + \{C(2) + C(12)\} \\
&= \{C(3) + C(12)\} + \{C(9) + C(2)\} + \{C(1) + C(4)\} + \{C(11) + C(10)\} \\
&= b + c + a + d = e + f = -\frac{1}{2}. \\
2cd &= 2\{C(2) + C(8)\} \cdot \{C(6) + C(7)\} \\
&= 2\{C(2)C(6) + C(2)C(7) + C(8)C(6) + C(8)C(7)\} \\
&= \{C(4) + C(8)\} + \{C(5) + C(9)\} + \{C(2) + C(14)\} + \{C(1) + C(15)\} \\
&= \{C(4) + C(1)\} + \{C(8) + C(2)\} + \{C(5) + C(14)\} + \{C(9) + C(15)\} \\
&= a + c + b + c = a + b + 2c.
\end{aligned}
$$

It follows that
$$
\begin{aligned}
ef &= (a+c)(b+d) = ab + ad + bc + cd \\
&= 2(a+b+c+d) = 2(e+f) = -1.
\end{aligned}
$$

Therefore, e and f are the roots of the quadratic equation
$$x^2 + \frac{1}{2}x - 1 = 0.$$

Now
$$
\begin{aligned}
f &= b + d = C(3) + C(5) + C(6) + C(7) \\
&= (\cos 3\theta + \cos 7\theta) + \cos 5\theta + \cos 6\theta \\
&= 2\cos 2\theta \cos 5\theta + \cos 5\theta + \cos 6\theta < 0,
\end{aligned}
$$

because
$$\cos 2\theta > 0 \quad \left(0 < 2\theta = \frac{4\pi}{17} < \frac{\pi}{2}\right);$$
$$\cos 5\theta < 0 \quad \left(\frac{\pi}{2} < 5\theta = \frac{10\pi}{17} < \pi\right);$$
$$\cos 6\theta < 0 \quad \left(\frac{\pi}{2} < 6\theta = \frac{12\pi}{17} < \pi\right).$$

Hence $f < 0$ and $e > 0$. So, we obtain
$$e = \frac{-1+\sqrt{17}}{4}, \quad f = \frac{-1-\sqrt{17}}{4}.$$

Furthermore, a and c are the roots of
$$x^2 - ex - \frac{1}{4} = 0,$$

and
$$a - c = (\cos\theta - \cos 2\theta) + (\cos 4\theta - \cos 8\theta) > 0;$$

hence $a > c$, and we have
$$\begin{aligned}
a &= \frac{1}{2}\left(e + \sqrt{e^2+1}\right) = \frac{-1+\sqrt{17}}{8} + \frac{\sqrt{(-1+\sqrt{17})^2+16}}{8} \\
&= \frac{-1+\sqrt{17}+\sqrt{34-2\sqrt{17}}}{8}, \\
c &= \frac{1}{2}\left(e - \sqrt{e^2+1}\right) \\
&= \frac{-1+\sqrt{17}-\sqrt{34-2\sqrt{17}}}{8}.
\end{aligned}$$

Similarly, b and d are the roots of
$$x^2 - fx - \frac{1}{4} = 0,$$

and we have
$$\begin{aligned}
b &= \frac{1}{2}\left(f + \sqrt{f^2+1}\right) = \frac{-1-\sqrt{17}}{8} + \frac{\sqrt{(-1-\sqrt{17})^2+16}}{8} \\
&= \frac{-1-\sqrt{17}+\sqrt{34+2\sqrt{17}}}{8}, \\
d &= \frac{-1-\sqrt{17}-\sqrt{34+2\sqrt{17}}}{8}.
\end{aligned}$$

Finally, because
$$C(1)C(4) = \frac{1}{2}(C(3) + C(5)) = \frac{b}{2},$$
$C(1)$ and $C(4)$ are the roots of
$$x^2 - ax + \frac{b}{2} = 0.$$

Therefore,
$$\begin{aligned}\cos\frac{2\pi}{17} &= \cos\theta = C(1) = \frac{a + \sqrt{a^2 - 2b}}{2} \\ &= \frac{-1 + \sqrt{17} + \sqrt{34 - 2\sqrt{17}}}{16} \\ &\quad + \frac{1}{8}\sqrt{17 + 3\sqrt{17} - \sqrt{34 - 2\sqrt{17}} - 2\sqrt{34 + 2\sqrt{17}}},\end{aligned}$$

where we used
$$\begin{aligned}a^2 &= \{C(1) + C(4)\}^2 = \frac{1}{2}\{C(0) + C(2)\} + \frac{1}{2}\{C(0) + C(8)\} + 2C(1)\cdot C(4) \\ &= 1 + \frac{1}{2}\{C(2) + C(8)\} + \{C(3) + C(5)\} = 1 + \frac{c}{2} + b; \\ a^2 - 2b &= 1 + \frac{c}{2} - b \\ &= 1 + \frac{-1 + \sqrt{17} - \sqrt{34 - 2\sqrt{17}}}{16} - \frac{-1 - \sqrt{17} + \sqrt{34 + 2\sqrt{17}}}{8} \\ &= \frac{15 + \sqrt{17} - \sqrt{34 - 2\sqrt{17}} + 2\left(1 + \sqrt{17}\right) - 2\sqrt{34 + 2\sqrt{17}}}{16} \\ &= \frac{17 + 3\sqrt{17} - \sqrt{34 - 2\sqrt{17}} - 2\sqrt{34 + 2\sqrt{17}}}{16}.\end{aligned}$$

This shows that $\cos\frac{2\pi}{17}$, hence a regular 17-gon too, are constructible.

By now, the reader has every right to be puzzled how do we know to group the roots of the equation
$$1 + z + z^2 + z^3 + \cdots + z^{16} = 0$$
so that the argument works perfectly? To answer this question, we need the following theorem from the number theory. (The proof can be found in many textbooks.)

Theorem. (i) For any prime number p, there exists r (called a *primitive root* (mod p)) such that the smallest positive integer m for which
$$r^m \equiv 1 \pmod{p}$$

Angle Trisection

is $m = p - 1$.

(ii) Suppose r is a primitive root (mod p). Then
$$1, r, r^2, \cdots, r^{p-1}$$
is a reduced system of residues.

We know, for any root ρ of the equation
$$1 + z + z^2 + \cdots + z^{16} = 0,$$
$$\rho, \rho^2, \rho^3, \cdots, \rho^{16}$$
represent all the roots. But the exponents do not have to be in arithmetic progression.

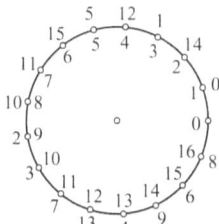

Figure ATS14

It is simple to check that 3 is a primitive root (mod 17), and so
$$3^0, 3^1, 3^2, \cdots, 3^{15};$$
i.e., (mod 17),
$$1, 3, 9, 10, 13, 5, 15, 11, 16, 14, 8, 7, 4, 12, 2, 6,$$
is a reduced system of residues. Thus, we denote

$\rho_0 = \omega^{3^0} = \omega^1$ $\qquad$ $\rho_1 = \rho_0^3 = \omega^3$
$\rho_2 = \rho_1^3 = \omega^9$ $\qquad$ $\rho_3 = \rho_2^3 = \omega^{27} = \omega^{10}$
$\rho_4 = \rho_3^3 = \omega^{13}$ $\qquad$ $\rho_5 = \rho_4^3 = \omega^5$
$\rho_6 = \rho_5^3 = \omega^{15}$ $\qquad$ $\rho_7 = \rho_6^3 = \omega^{11}$
$\rho_8 = \rho_7^3 = \omega^{16} \, (= \omega^{-1})$ $\qquad$ $\rho_9 = \rho_8^3 = \omega^{14}$
$\rho_{10} = \rho_9^3 = \omega^8$ $\qquad$ $\rho_{11} = \rho_{10}^3 = \omega^7$
$\rho_{12} = \rho_{11}^3 = \omega^4$ $\qquad$ $\rho_{13} = \rho_{12}^3 = \omega^{12}$
$\rho_{14} = \rho_{13}^3 = \omega^2$ $\qquad$ $\rho_{15} = \rho_{14}^3 = \omega^6$

Note that if we continue, then we obtain
$$\rho_{16} = \rho_{15}^3 = \omega = \rho_0,$$

and we'll be simply repeating the cycle (i.e., the subscripts are modulo 16).

Taking alternate terms, let

$$E = \rho_0 + \rho_2 + \rho_4 + \rho_6 + \rho_8 + \rho_{10} + \rho_{12} + \rho_{14},$$
$$F = \rho_1 + \rho_3 + \rho_5 + \rho_7 + \rho_9 + \rho_{11} + \rho_{13} + \rho_{15}.$$

We claim that E and F are the roots of a quadratic equation with rational (in fact, integer) coefficients. We know

$$E + F = \rho_0 + \rho_1 + \rho_2 + \cdots + \rho_{15} = -1.$$

So all we need is to compute the product

$$EF = (\rho_0 + \rho_2 + \cdots + \rho_{14})(\rho_1 + \rho_3 + \cdots + \rho_{15}).$$

But the product can be expanded into the sum of 64 terms of the form $\rho_i \rho_j$, where none of the terms equals to 1 (because the multiplicative inverse of ρ_i is ρ_{i+8}, and these two are either both in E, or both in F), hence this product can be rewritten as a linear combination of $\rho_0, \rho_1, \cdots, \rho_{15}$; i.e.,

$$EF = c_0 \rho_0 + c_1 \rho_1 + c_2 \rho_2 + \cdots + c_{15} \rho_{15},$$

where all the coefficients are nonnegative integers. (Why?) Note that the sum of all the coefficients must be 64.

Now all ρ's are defined in terms of ρ_0; i.e., $\rho_i = \rho_0^{3^i}$. Thus, if we substitute ρ_1 for ρ_0, then ρ_i becomes

$$\rho_1^{3^i} = \left(\rho_0^{3^1}\right)^{3^i} = \rho_0^{3^{i+1}} = \rho_{i+1}.$$

(That is, the subscripts are shifted by 1 (mod 16)). Hence E becomes F, and vice versa. It follows that the product EF is invariant under the substitution ρ_1 for ρ_0, implying that

$$c_0 = c_1 = c_2 = \cdots = c_{15}; \text{ i.e., } EF = c_0(\rho_0 + \rho_1 + \cdots + \rho_{15}) = -c_0.$$

But the product expands into the sum of 64 terms, and so we conclude that

$$c_0 = 4 \quad \text{and} \quad EF = -4.$$

We have shown that E and F are the roots of the quadratic equation with integer coefficients:

$$X^2 + X - 4 = 0.$$

Next, again taking alternate terms from E and F, let

$$A = \rho_0 + \rho_4 + \rho_8 + \rho_{12}, \quad B = \rho_1 + \rho_5 + \rho_9 + \rho_{13},$$
$$C = \rho_2 + \rho_6 + \rho_{10} + \rho_{14}, \quad D = \rho_3 + \rho_7 + \rho_{11} + \rho_{15}.$$

Angle Trisection

We have
$$A + C = E,$$
hence to find the quadratic equation satisfied by A and C, all we need is to find the product
$$AC = (\rho_0 + \rho_4 + \rho_8 + \rho_{12})(\rho_2 + \rho_6 + \rho_{10} + \rho_{14}).$$
Again this product can be expanded into the sum of 16 terms of the form $\rho_i \rho_j$, where none of the terms equals 1, and so this product can be rewritten as a linear combination of ρ's; i.e.,
$$AC = c_0 \rho_0 + c_1 \rho_1 + c_2 \rho_2 + \cdots + c_{15} \rho_{15}.$$
Now if we substitute ρ_2 for ρ_0 in the definition of ρ's, then $\rho_i = \rho_0^{3^i}$ becomes
$$\rho_2^{3^i} = \left(\rho_0^{3^2}\right)^{3^i} = \rho_0^{3^{i+2}} = \rho_{i+2};$$
i.e., the subscripts are shifted by 2 (mod 16), and so A becomes C, and vice versa. This means that the product AC is invariant under the substitution ρ_2 for ρ_0, implying that
$$c_0 = c_2 = \cdots = c_{14}, \quad c_1 = c_3 = \cdots = c_{15}.$$
It follows that
$$\begin{aligned} AC &= c_0(\rho_0 + \rho_2 + \cdots + \rho_{14}) + c_1(\rho_1 + \rho_3 + \cdots + \rho_{15}) \\ &= c_0 E + c_1 F. \end{aligned}$$

Considering that the product expands into 16 terms, we must have one of the following three cases:
$$\begin{array}{ll} AC = 2E & (c_0 = 2,\ c_1 = 0); \\ AC = 2F & (c_0 = 0,\ c_1 = 2); \\ AC = E + F & (c_0 = c_1 = 1). \end{array}$$
But
$$\rho_0 \rho_2 = \omega \cdot \omega^9 = \omega^{10} = \rho_3 \in F \quad \text{and} \quad \rho_0 \rho_6 = \omega \cdot \omega^{15} = \omega^{16} = \rho_8 \in E.$$
Therefore, none of the first two cases can happen; i.e.,
$$AC = E + F = -1.$$
We have established that A and C are the roots of the quadratic equation
$$X^2 - EX - 1 = 0.$$
Similarly, B and D are the roots of the quadratic equation
$$X^2 - FX - 1 = 0.$$

(This assertion can also be obtained by substituting ρ_1 for ρ_0 in the definition of the ρ's. Then A, C, and E, become B, D, and F, respectively.)

Note that (by taking alternate terms), we have

$$\begin{aligned}
A &= (\rho_0 + \rho_8) + (\rho_4 + \rho_{12}) = 2(\cos\theta + \cos 4\theta), \\
B &= (\rho_1 + \rho_9) + (\rho_5 + \rho_{13}) = 2(\cos 3\theta + \cos 5\theta), \\
C &= (\rho_2 + \rho_{10}) + (\rho_6 + \rho_{14}) = 2(\cos 8\theta + \cos 2\theta), \\
D &= (\rho_3 + \rho_{11}) + (\rho_7 + \rho_{15}) = 2(\cos 7\theta + \cos 6\theta).
\end{aligned}$$

Furthermore,

$$(2\cos\theta)(2\cos 4\theta) = (\rho_0 + \rho_8)(\rho_4 + \rho_{12}).$$

The product on the right can be expanded as the sum of four terms of the form $\rho_i \rho_j$, where none of the term equals to 1. And if we substitute ρ_4 for ρ_0 in the definition of the ρ's, then $(\rho_0 + \rho_8)$ becomes $(\rho_4 + \rho_{12})$, and vice versa. Therefore, this product must be equal to one of A, B, C, or D. But

$$\rho_0 \rho_4 = \omega\omega^{13} = \omega^{14} = \rho_9 \in B.$$

It follows that

$$(2\cos\theta)(2\cos 4\theta) = B.$$

We have shown that $2\cos\theta$ and $2\cos 4\theta$ are the roots of the quadratic equation

$$X^2 - AX + B = 0.$$

Similarly,

$$(2\cos 3\theta)(2\cos 5\theta) = (\rho_1 + \rho_9)(\rho_5 + \rho_{13}) = C.$$

And so $2\cos 3\theta$ and $2\cos 5\theta$ are the roots of the quadratic equation

$$X^2 - BX + C = 0.$$

Moreover,

$$(2\cos 8\theta)(2\cos 2\theta) = (\rho_2 + \rho_{10})(\rho_6 + \rho_{14}) = D.$$

And so $2\cos 8\theta$ and $2\cos 2\theta$ are the roots of the quadratic equation

$$X^2 - CX + D = 0.$$

Finally,

$$(2\cos 7\theta)(2\cos 6\theta) = (\rho_3 + \rho_{11})(\rho_7 + \rho_{15}) = A.$$

And so $2\cos 7\theta$ and $2\cos 6\theta$ are the roots of the quadratic equation

$$X^2 - DX + A = 0.$$

From $\theta = \frac{2\pi}{17}$, $\omega = \cos\theta + i\sin\theta$, we obtain

$$\cos\theta > \cos 2\theta > \cos 3\theta > \cos 4\theta > 0 > \cos 5\theta > \cos 6\theta > \cos 7\theta > \cos 8\theta,$$

Angle Trisection

implying that

$$\begin{aligned}
A &= 2(\cos\theta + \cos 4\theta) &&> 0; \\
B &= 2(\cos 3\theta + \cos 5\theta) &= 4\cos\theta\cos 4\theta &> 0; \\
C &= 2(\cos 2\theta + \cos 8\theta) &= 4\cos 5\theta\cos 3\theta &< 0; \\
D &= 2(\cos 6\theta + \cos 7\theta) &&< 0.
\end{aligned}$$

Furthermore,

$$A = 4\cos\frac{5\theta}{2}\cos\frac{3\theta}{2} > 4|\cos 5\theta|\cos 3\theta = |C|$$

(because $\cos\frac{5\theta}{2} > |\cos 5\theta|$ and $\cos\frac{3\theta}{2} > \cos 3\theta$). It follows that

$$E = A + C > 0, \quad A > C, \quad F = B + D < 0, \quad B > D.$$

Therefore, we can determine all the signs before the square root signs. We have

$$\begin{aligned}
E &= \tfrac{1}{2}\{-1+\sqrt{17}\}, & F &= \tfrac{1}{2}\{-1-\sqrt{17}\}, \\
A &= \tfrac{1}{2}\{E+\sqrt{E^2+4}\}, & B &= \tfrac{1}{2}\{F+\sqrt{F^2+4}\}, \\
C &= \tfrac{1}{2}\{E-\sqrt{E^2+4}\}, & D &= \tfrac{1}{2}\{F-\sqrt{F^2+4}\}, \\
2\cos\theta &= \tfrac{1}{2}\{A+\sqrt{A^2-4B}\}, & 2\cos 2\theta &= \tfrac{1}{2}\{C+\sqrt{C^2-4D}\}, \\
2\cos 3\theta &= \tfrac{1}{2}\{B+\sqrt{B^2-4C}\}, & 2\cos 4\theta &= \tfrac{1}{2}\{A-\sqrt{A^2-4B}\}, \\
2\cos 5\theta &= \tfrac{1}{2}\{B-\sqrt{B^2-4C}\}, & 2\cos 6\theta &= \tfrac{1}{2}\{D+\sqrt{D^2-4A}\}, \\
2\cos 7\theta &= \tfrac{1}{2}\{D-\sqrt{D^2-4A}\}, & 2\cos 8\theta &= \tfrac{1}{2}\{C-\sqrt{C^2-4D}\}.
\end{aligned}$$

Note that all these values are real, and if we substitute all the numerical values, then we obtain three layers of square root signs for the values of cosines.

Exercise. Construct angle $\theta = \frac{2\pi}{17}$ with straightedge and compass.

Theorem. [Gauss] Suppose p is an odd prime, then a regular p-gon is constructible (if and) only if $p = 2^m + 1$.

Proof. Recall that $\omega = \cos\frac{2\pi}{p} + i\sin\frac{2\pi}{p}$ is a root of

$$x^{p-1} + x^{p-2} + \cdots + x + 1 = 0,$$

and this polynomial equation of degree $(p-1)$ is irreducible over the rational field. (Substitute $x = y+1$ and apply the Eisenstein theorem.) Hence a regular p-gon is constructible if and only if $\cos\frac{2\pi}{p}$ can be obtained by a finite number of rational operations and extracting square roots starting from the rational field. It follows that for a regular p-gon to be constructible, we must have $p - 1 = 2^m$. □

Remark. Conversely, if $p = 2^m + 1$ is a prime number, then a regular p-gon, in fact, is constructible. Furthermore, for $2^m + 1$ to be a prime, m can not have an odd factor. (Why?) Hence we must have $m = 2^k$ for some integer k. Fermat asserted that all the numbers of this form are primes. (Integers of this form are called *Fermat*

numbers.) His assertion is correct for the first five of them: for $k = 0, 1, 2, 3, 4$, we have
$$p = 3, 5, 17, 257, 65537,$$
all of them are primes. However, in 1732, L. Euler showed that the next one is divisible by 641; i.e., for $k = 5$,
$$p = 2^{32} + 1 = 4294967297 = 641 \times 6700417.$$

In fact, no further prime of this form is known.

Chapter 21

Conics

Inspired by the intriguing Japanese book by NAKAZAWA Sadaharu, *From the Mathematics Classroom Window* with the subtitle, *The Backstage of Composing Mathematics Problems* (Gendai Sugaku Sha, Kyoto, 1990), we shall introduce some beautiful, yet not very well-known theorems on conics in this chapter.

Recall that the *orthocenter* of a triangle is the intersection of its three altitudes[1], and a hyperbola is *equilateral* if (and only if) its two asymptotes are perpendicular to each other.

Theorem 1. Suppose a triangle has all three of its vertices on an equilateral hyperbola. Then its orthocenter is also on the equilateral hyperbola.

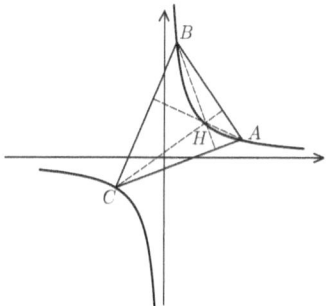

Figure CNC1

Proof. Without loss of generality, we may assume that the equation of the equilateral hyperbola is $xy = 1$ (Figure CNC1), and the three vertices of the triangle are

$$A\left(a, \frac{1}{a}\right), \quad B\left(b, \frac{1}{b}\right), \quad C\left(c, \frac{1}{c}\right) \quad (abc \neq 0).$$

---

[1] See my book, *Complex Numbers and Geometry* (Mathematical Association of America, 1994), Appendix A, pp 171 - 172.

Then the slope of line BC is

$$\frac{\frac{1}{c} - \frac{1}{b}}{c - b} = -\frac{1}{bc}.$$

Hence the equation of the altitude from vertex A (perpendicular to BC) is

$$y - \frac{1}{a} = bc(x - a).$$

Solving

$$bcx = -\frac{1}{a}, \quad y = -abc,$$

we see that point

$$H\left(-\frac{1}{abc}, -abc\right)$$

is on the altitude from vertex A.

But the coordinates of point $H(-\frac{1}{abc}, -abc)$ are symmetric with respect to a, b, c; and so point H is also on the altitude from B (and also on the altitude from C). Therefore, $H(-\frac{1}{abc}, -abc)$ is the orthocenter of $\triangle ABC$, and clearly H is on our equilateral hyperbola $xy = 1$. □

Note that each of the four points A, B, C, H, is the orthocenter of the triangle with the remaining three as its vertices. Such a system of four points is called an *orthocentric system*. Furthermore, the four circumradii of the triangles with their vertices in an orthocentric system are equal (though the circumcircles are different). Note also that one of the points in an orthocentric system must be on the other branch of the equilateral hyperbola.

Exercise. Show that, given three non-collinear points, there exists an equilateral hyperbola passing through the three given points. How many such equilateral hyperbolas are there?

> *Theorem 2.* Let P and Q be a pair of points on an equilateral hyperbola symmetric with respect to the center O of the hyperbola. Draw a circle centered at P passing through point Q. Then, of the four intersections of the circle and the equilateral hyperbola, three of them (other than point Q) are the vertices of an equilateral triangle. (Figure CNC2) Naturally, the three intersections (other than point P), of the hyperbola and the circle centered at Q passing through point P, are also the vertices of an equilateral triangle. Furthermore, these two equilateral triangles are in orthogonal position.[2]

---

[2] See my book, *Honsberger Revisited* (National Taiwan University Press, 2012), page 1.

Conics

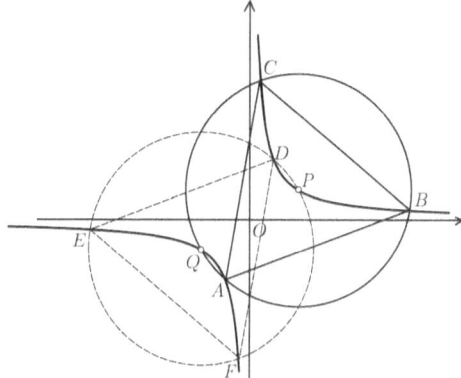

Figure CNC2

Proof. We use complex numbers. With

$$x = \frac{z + \bar{z}}{2}, \quad y = \frac{z - \bar{z}}{2i},$$

the equation $xy = 1$ of the equilateral hyperbola becomes

$$z^2 - \bar{z}^2 = 4i.$$

Let z_0 be the complex number representing point P on the hyperbola. A point z on the circle centered at P with radius $\overline{PQ} = 2\overline{OP}$ can be represented as

$$z = z_0 + 2z_0\zeta \quad \text{(where } \zeta \text{ is a parameter satisfying } |\zeta| = 1).$$

To find the condition that this point z is also on the equilateral hyperbola, we compute

$$\begin{aligned}
z^2 - \bar{z}^2 &= z_0^2(1 + 2\zeta)^2 - \bar{z}_0^2(1 + 2\bar{\zeta})^2 \\
&= (z_0^2 - \bar{z}_0^2) + 4z_0^2(\zeta + \zeta^2) - 4\bar{z}_0^2(\bar{\zeta} + \bar{\zeta}^2).
\end{aligned}$$

Because point $P(z_0)$ is on the equilateral hyperbola, we have

$$z_0^2 - \bar{z}_0^2 = 4i.$$

Therefore, a necessary and sufficient condition for point $z = z_0 + 2z_0\zeta$ ($|\zeta| = 1$) to be on the equilateral hyperbola is

$$z_0^2(\zeta + \zeta^2) - \bar{z}_0^2(\bar{\zeta} + \bar{\zeta}^2) = 0.$$

Because $|\zeta| = 1$, using $\bar{\zeta} = \frac{1}{\zeta}$, this condition becomes

$$(\zeta + 1)\left(z_0^2 \zeta - \frac{\bar{z}_0^2}{\zeta^2}\right) = 0.$$

Now $\zeta = -1$ gives point Q (that is symmetric to P with respect to the center O of the equilateral hyperbola), and the remaining three roots ζ_1, ζ_2, ζ_3, satisfy

$$\zeta^3 = \left(\frac{\bar{z}_0}{z_0}\right)^2 ; \text{ i.e., } \zeta_k = \left(\frac{\bar{z}_0}{z_0}\right)^{2/3} \omega^k \quad (k = 0, 1, 2),$$

where ω is a primitive cube root of 1. Clearly, these roots give three points

$$z_1 = z_0 + 2z_0\zeta_1, \quad z_2 = z_0 + 2z_0\zeta_2, \quad z_3 = z_0 + 2z_0\zeta_3,$$

that are the vertices of an equilateral triangle (inscribed in circle $|z - z_0| = 2|z_0|$ and on the equilateral hyperbola).[3]

Obviously, the two equilateral triangles are symmetric with respect to the center O of the equilateral hyperbola, hence they are in orthogonal position. □

Remarks. (a) Had we used the equation $x^2 - y^2 = a^2$ for our equilateral hyperbola, the computation will be almost identical. (Readers are encouraged to give it a try.)
(b) It turns out that the radius $= 2\overline{OP}$, is also a necessary condition; i.e., no other radius gives three points that are the vertices of an equilateral triangle and on both the circle and the equilateral hyperbola.

Theorem 3. Let $\triangle ABC$ be formed by the tangents at three distinct points U, V, W, on a parabola. Then

(a) the orthocenter H of $\triangle ABC$ is on the directrix of the parabola.

(b) $[ABC] = -\frac{1}{2}[UVW]$.

(Note that we are considering the signed areas.)

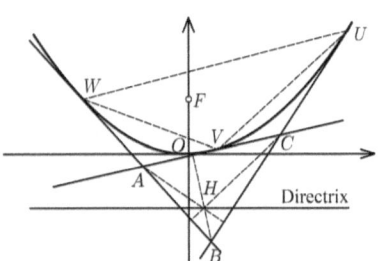

Figure CNC3

Proof. (a) Let the equation of the parabola be $4py = x^2$ (Figure CNC2), and the three points on the parabola be

$$U(2pu, pu^2), \quad V(2pv, pv^2), \quad W(2pw, pw^2).$$

---
[3] Readers who would like to learn the technique of using complex numbers will find much to enjoy in my book, *Complex Numbers and Geometry* (Mathematical Association of America, 1994).

Conics

Because[4] the equation of the tangent at point (x_0, y_0) on parabola $4py = x^2$ is

$$2p(y + y_0) = x_0 x;$$

the equation of the tangent at point $U(2pu, pu^2)$ must be

$$2p(y + pu^2) = 2pux; \text{ i.e., } y = ux - pu^2.$$

(Thus the slope of the altitude from vertex A must be $-\frac{1}{u}$.) Similarly, the equations of the tangents at V and W are, respectively,

$$y = vx - pv^2 \quad \text{and} \quad y = wx - pw^2.$$

Solving the system of simultaneous equations consisting of the last two equations, we obtain the coordinates of vertex A (of $\triangle ABC$) as

$$A(p(v+w), pvw).$$

Hence the equation of the altitude from vertex A is

$$-u(y - pvw) = x - p(v+w); \text{ i.e., } x + uy = p\{uvw + (v+w)\}.$$

The intersection of this altitude and the directrix $y + p = 0$ is

$$H(p\{uvw + (u+v+w)\}, -p).$$

The coordinates of this point H is symmetric with respect to u, v, w; and so the altitude from vertex B must also pass through this point (so does the altitude from vertex C). Hence this point H that is on the directrix $y + p = 0$ must be the orthocenter of $\triangle ABC$.

(b) Using the results of the computations above, we have

$$[ABC] = \frac{1}{2} \begin{vmatrix} p(v+w) & pvw & 1 \\ p(w+u) & pwu & 1 \\ p(u+v) & puv & 1 \end{vmatrix} = \frac{p^2}{2} \begin{vmatrix} v+w & vw & 1 \\ w+u & wu & 1 \\ u+v & uv & 1 \end{vmatrix}.$$

Clearly, this determinant (if expanded) is a homogeneous cubic polynomial in u, v, w. And alternate in the sense that if any two of the variables are equal, then (two rows will be identical, and so) the determinant vanishes. Therefore, for some constant k, we must have

$$\begin{vmatrix} v+w & vw & 1 \\ w+u & wu & 1 \\ u+v & uv & 1 \end{vmatrix} = k(v-w)(w-u)(u-v).$$

---
[4] See my book, *Honsberger Revisited* (National Taiwan University Press, Taipei, 2012), Appendix C: *Useful Theorems*.

Considering each side as a quadratic polynomial in u, and comparing the "coefficients" of the u^2-term of both sides, we obtain $k = -1$. Therefore,

$$[ABC] = -\frac{p^2}{2}(v-w)(w-u)(u-v).$$

On the other hand,

$$[UVW] = \frac{1}{2}\begin{vmatrix} 2pu & pu^2 & 1 \\ 2pv & pv^2 & 1 \\ 2pw & pw^2 & 1 \end{vmatrix} = p^2 \begin{vmatrix} 1 & u & u^2 \\ 1 & v & v^2 \\ 1 & w & w^2 \end{vmatrix}$$
$$= p^2(v-w)(w-u)(u-v).$$

The last determinant is known as the *Vandermonde determinant*. (Those who are not familiar with the Vandermonde determinants may imitate the argument employed in factoring the determinant for $[ABC]$.) We have shown that

$$[ABC] = -\frac{1}{2}[UVW].$$

□

Exercise. Suppose two tangents that touch parabola $4py = x^2$ at U and V, respectively, intersect at point P on line $y + 2p = 0$. Show that the orthocenter of $\triangle PUV$ is at the vertex of the parabola.

Theorem 4. Inscribe a right triangle in a parabola with the right-angled vertex at the vertex of the parabola. Then the hypotenuse must pass through a (common) fixed point.

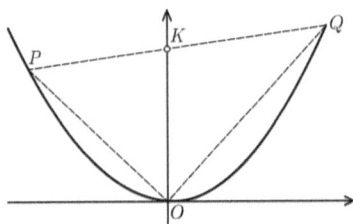

Figure CNC4

Proof. Let PQ (where $P(2pu, pu^2)$, $Q(2pv, pv^2)$) be the hypotenuse of a right triangle inscribed in the parabola $4py = x^2$. Suppose there exists a fixed point as asserted. Then, by symmetry, this fixed point must be on the axis of the parabola. And by considering the isosceles right triangle inscribed in the parabola, we conjecture that the fixed point must be $K(0, 4p)$. (Figure CNC4)

Conics

To show that points P, Q, and K are collinear, we compute the slopes. The slope of PK is
$$\frac{p(u^2-4)}{2pu} = \frac{u}{2} - \frac{2}{u};$$
and similarly, that of QK is
$$\frac{v}{2} - \frac{2}{v}.$$
We want to show that these two slopes are equal if $OP \perp OQ$. In this case, we have
$$\frac{pu^2}{2pu} \cdot \frac{pv^2}{2pv} = -1; \text{ i.e., } \frac{u}{2} \cdot \frac{v}{2} = -1;$$
and so $\frac{v}{2} = -\frac{2}{u}$, implying the desired result. □

Theorem 5. Let P and Q be points on ellipse
$$\frac{x^2}{a^2} + \frac{y^2}{b^2} = 1$$
such that $OP \perp OQ$, where O is the center of the ellipse. (Figure CNC5) Then chord PQ is tangent to a fixed circle.

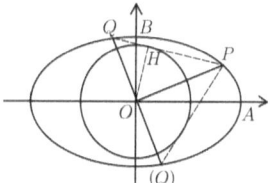

Figure CNC5

Proof. Let
$$A(a, 0), \quad B(0, b), \quad \overline{OP} = p, \quad \overline{OQ} = q, \quad \angle AOP = \theta.$$
Then we have $P(p\cos\theta, p\sin\theta)$, and
$$Q\left(q\cos(\theta \pm \frac{\pi}{2}), q\sin(\theta \pm \frac{\pi}{2})\right) = Q(\mp q\sin\theta, \pm q\cos\theta).$$
Furthermore, let H be the foot of the perpendicular from O to chord PQ. Then
$$[OPQ] = \frac{1}{2}\overline{OH} \cdot \overline{PQ} = \frac{1}{2}\overline{OP} \cdot \overline{OQ}.$$
Therefore,
$$\overline{OH} = \frac{\overline{OP} \cdot \overline{OQ}}{\overline{PQ}} = \frac{pq}{\sqrt{p^2+q^2}}.$$

Because $P(p\cos\theta, p\sin\theta)$ is on the ellipse, we have

$$\left(\frac{p\cos\theta}{a}\right)^2 + \left(\frac{p\sin\theta}{b}\right)^2 = 1, \quad p^2 = \frac{a^2b^2}{a^2\sin^2\theta + b^2\cos^2\theta}.$$

Similarly,
$$q^2 = \frac{a^2b^2}{a^2\cos^2\theta + b^2\sin^2\theta}.$$

Hence
$$p^2 + q^2 = a^2b^2\left(\frac{1}{a^2\sin^2\theta + b^2\cos^2\theta} + \frac{1}{a^2\cos^2\theta + b^2\sin^2\theta}\right)$$
$$= \frac{a^2b^2(a^2+b^2)}{(a^2\sin^2\theta + b^2\cos^2\theta)\cdot(a^2\cos^2\theta + b^2\sin^2\theta)}.$$

Substituting these values, we obtain

$$\overline{OH}^2 = \frac{p^2q^2}{p^2+q^2} = \frac{a^2b^2}{a^2+b^2},$$

which is independent of the points P and Q. □

Note that, by the A.M.-G.M. inequality, we have

$$\overline{OH} \leq \frac{\sqrt{a^2+b^2}}{2},$$

i.e., $\overline{OH}$ can not be greater than the median of the right triangle OAB from vertex O, which should be obvious geometrically. Equality holds if and only if the ellipse is a circle.

Chapter 22

Primes

22.1 Number of Primes

When considering the divisibility, the sign of an integer does not matter, so we shall consider only nonnegative integers in this chapter. In terms of the number of its divisors, integers can be divided into four classes.

First, 0 is divisible by any non-zero integers, so it has infinitely many divisors.

Next, 1 has no divisor other than 1 itself; i.e., 1 has only one divisor. Note that 0 and 1 are not in the same class.

Now an integer $a > 1$ has at least two divisors; 1 and a itself. A divisor of a other than 1 and a is called a *proper divisor* (of a). An integer that has a proper divisor is called a *composite* number. For example, 4, 6, 8, 9, 10, are composite numbers.

An integer that has no proper divisor is called a *prime* or a *prime number*. For example, 2, 3, 5, 7, 11, 13, 17, are primes.

Summing up, ignoring the signs, integers can be divided into four classes:

0	:	Infinitely many divisors
1 (unit)	:	Only one divisor
Composite Numbers	:	Have proper divisors
Prime Numbers	:	No proper divisor

Note that 1, which may be considered as the prime among the primes, is excluded from the class of primes because otherwise almost all the statements of theorems in number theory will become cumbersome.

The very first and natural question about primes is that how many primes are there? Are there finitely many or infinitely many primes? This question is elegantly answered by Euclid. Let primes be denoted in increasing order:

$$p_1 = 2, \ p_2 = 3, \ p_3 = 5, \ p_4 = 7, \ p_5 = 11, \ p_6 = 13, \ p_7 = 17, \ p_8 = 19, \ \cdots.$$

Suppose there are only k of them. We multiply all of them and add 1; i.e., consider

$$m = p_1 \cdot p_2 \cdot p_3 \cdots p_k + 1.$$

Here are the first few:

$$2 + 1 = 3$$
$$2 \cdot 3 + 1 = 7$$
$$2 \cdot 3 \cdot 5 + 1 = 31$$
$$2 \cdot 3 \cdot 5 \cdot 7 + 1 = 211$$
$$2 \cdot 3 \cdot 5 \cdot 7 \cdot 11 + 1 = 2311$$
$$2 \cdot 3 \cdot 5 \cdot 7 \cdot 11 \cdot 13 + 1 = 30031 = 59 \cdot 509$$

Clearly, m is bigger than any of the primes. Now, m must be either a prime or a composite. If m is a prime, then we find a prime other than p_1, p_2, $\cdots$, p_k, which contradicts our assumption that there are no others. Suppose m is a composite number. Then because it is divisible by none of the primes p_1, p_2, $\cdots$, p_k, so it must have a prime divisor other than these, which again establishes the existence of a new prime. In either case, we obtain a contradiction if we assume that there are only finitely many primes. Hence we must conclude that there are infinitely many primes.

Now, obviously, 2 is the only even prime. All other primes are odd. Odd primes can be further classified into two classes: those that give remainder 1, and those that give remainder 3 (or, if you wish, remainder -1), when divided by 4. They can be represented as $4n + 1$ and $4n - 1$, respectively. Primes 5, 13, 17, 29 are examples of the first class, while primes 3, 7, 11, 19, 23 are that of the second class. Again we can raise the question whether either class contains only finitely many primes? For primes of the form $4n - 1$, we can imitate Euclid's reasoning. Suppose there are only finitely many primes of the form $4n - 1$, say q_1, q_2, $\cdots$, q_k. Consider

$$m = 4(q_1 \cdot q_2 \cdots q_k) - 1.$$

If this number m is a prime, then it is certainly a new prime of the form $4n - 1$, and we obtain a contradiction. If this number is composite, then it must have a prime divisor of the form $4n - 1$ because no matter how many primes of the form $4n + 1$ are multiplied together, we will not obtain a number of the form $4n - 1$. But none of the primes q_1, q_2, $\cdots$, q_k divides our number m. So there must be a new prime of the form $4n-1$, implying that there must be infinitely many primes of the form $4n-1$.

Exercise. (a) Can we prove the existence of infinitely many primes of the form $4n+1$ by imitating Euclid's reasoning? If not, why not?
(b) How about the primes of the form $3n + 1$ or $3n - 1$?

Another application of Euclid's reasoning is that we can use it to prove that a gap between two primes can be as large as we please. For example, if we want to show that there are 100 consecutive integers none of them is a prime, all we have to do is to choose an integer k greater than 100 (say, $k = 101$), and consider the numbers

$$k! + 2,\ k! + 3,\ k! + 4,\ \cdots,\ k! + 101,$$

Primes

where $k!$ denotes the product of all the positive integers not greater than k. For example, $5! = 1 \cdot 2 \cdot 3 \cdot 4 \cdot 5$. Then the first number is divisible by 2, the second by 3, the third by 4, etc., and so none of these 100 consecutive integers is a prime.

Exercise. Show that for $n = 1, 2, \cdots, 16$, the values of function

$$f(n) = n^2 - n + 17$$

are all primes.

22.2 An Open Problem

Euclid's proof of the infinitude of primes works even if we split the set of primes into two disjoint sets. Let P_n be the set of the first n primes; i.e.,

$$P_n = \{p_1, p_2, p_3, \cdots, p_n\},$$

where $p_1 = 2$, $p_2 = 3$, $p_3 = 5$, etc. Let E be a subset of P_n, and consider

$$\prod_{p_j \in E} p_j + \prod_{p_k \in P_n \setminus E} p_k.$$

We adopt the convention that

$$\prod_{p_k \in \emptyset} p_k = 1.$$

(In the original proof by Euclid, $E = P_n$ and so $P_n \setminus E = \emptyset$.) Then, for P_2, we have

$$2 \cdot 3 + 1 = 7, \quad 2 + 3 = 5.$$

Both are primes. For P_3, we have

$$2 \cdot 3 \cdot 5 + 1 = 31, \quad 2 \cdot 3 + 5 = 11, \quad 2 \cdot 5 + 3 = 13, \quad 3 \cdot 5 + 2 = 17.$$

All of them are primes. For P_4, we have

$$
\begin{aligned}
2 \cdot 3 \cdot 5 \cdot 7 + 1 &= 211 & 2 \cdot 3 + 5 \cdot 7 &= 41 \\
2 \cdot 3 \cdot 5 + 7 &= 37 & 2 \cdot 5 + 3 \cdot 7 &= 31 \\
2 \cdot 3 \cdot 7 + 5 &= 47 & 2 \cdot 7 + 3 \cdot 5 &= 29 \\
2 \cdot 5 \cdot 7 + 3 &= 73 \\
3 \cdot 5 \cdot 7 + 2 &= 107
\end{aligned}
$$

Again all of them are primes. However, for P_5, we have

$2 \cdot 3 \cdot 5 \cdot 7 \cdot 11 + 1$	=	2311		$2 \cdot 3 \cdot 5 + 7 \cdot 11$	=	107	
$2 \cdot 3 \cdot 5 \cdot 7 + 11$	=	221	= $13 \cdot 17$	$2 \cdot 3 \cdot 7 + 5 \cdot 11$	=	97	
$2 \cdot 3 \cdot 5 \cdot 11 + 7$	=	337		$2 \cdot 5 \cdot 7 + 3 \cdot 11$	=	103	
$2 \cdot 3 \cdot 7 \cdot 11 + 5$	=	467		$3 \cdot 5 \cdot 7 + 2 \cdot 11$	=	127	
$2 \cdot 5 \cdot 7 \cdot 11 + 3$	=	773		$2 \cdot 3 \cdot 11 + 5 \cdot 7$	=	101	
$3 \cdot 5 \cdot 7 \cdot 11 + 2$	=	1157	= $13 \cdot 89$	$2 \cdot 5 \cdot 11 + 3 \cdot 7$	=	131	
				$3 \cdot 5 \cdot 11 + 2 \cdot 7$	=	179	
				$2 \cdot 7 \cdot 11 + 3 \cdot 5$	=	169	= 13^2
				$3 \cdot 7 \cdot 11 + 2 \cdot 5$	=	241	
				$5 \cdot 7 \cdot 11 + 2 \cdot 3$	=	391	= $17 \cdot 23$

Here only four of them fail to be primes out of 16. Now, for P_6 we have

$$
\begin{aligned}
2 \cdot 3 \cdot 5 \cdot 7 \cdot 11 \cdot 13 + 1 &= 30031 = 59 \cdot 509 \\
2 \cdot 3 \cdot 5 \cdot 7 \cdot 11 + 13 &= 2323 = 23 \cdot 101 \\
2 \cdot 3 \cdot 5 \cdot 7 \cdot 13 + 11 &= 2741 \\
2 \cdot 3 \cdot 5 \cdot 11 \cdot 13 + 7 &= 4297 \\
2 \cdot 3 \cdot 7 \cdot 11 \cdot 13 + 5 &= 6011 \\
2 \cdot 5 \cdot 7 \cdot 11 \cdot 13 + 3 &= 10013 = 17 \cdot 19 \cdot 31 \\
3 \cdot 5 \cdot 7 \cdot 11 \cdot 13 + 2 &= 15017 \\
2 \cdot 3 \cdot 5 \cdot 7 + 11 \cdot 13 &= 353 \\
2 \cdot 3 \cdot 5 \cdot 11 + 7 \cdot 13 &= 421 \\
2 \cdot 3 \cdot 7 \cdot 11 + 5 \cdot 13 &= 527 = 17 \cdot 31 \\
2 \cdot 5 \cdot 7 \cdot 11 + 3 \cdot 13 &= 809 \\
3 \cdot 5 \cdot 7 \cdot 11 + 2 \cdot 13 &= 1181 \\
2 \cdot 3 \cdot 5 \cdot 13 + 7 \cdot 11 &= 467 \\
2 \cdot 3 \cdot 7 \cdot 13 + 5 \cdot 11 &= 601 \\
2 \cdot 5 \cdot 7 \cdot 13 + 3 \cdot 11 &= 943 = 23 \cdot 41 \\
3 \cdot 5 \cdot 7 \cdot 13 + 2 \cdot 11 &= 1387 = 19 \cdot 73 \\
2 \cdot 3 \cdot 11 \cdot 13 + 5 \cdot 7 &= 893 = 19 \cdot 47 \\
2 \cdot 5 \cdot 11 \cdot 13 + 3 \cdot 7 &= 1451 \\
3 \cdot 5 \cdot 11 \cdot 13 + 2 \cdot 7 &= 2159 = 17 \cdot 127 \\
2 \cdot 7 \cdot 11 \cdot 13 + 3 \cdot 5 &= 2017 \\
3 \cdot 7 \cdot 11 \cdot 13 + 2 \cdot 5 &= 3013 = 23 \cdot 131 \\
5 \cdot 7 \cdot 11 \cdot 13 + 2 \cdot 3 &= 5011 \\
2 \cdot 3 \cdot 5 + 7 \cdot 11 \cdot 13 &= 1031 \\
2 \cdot 3 \cdot 7 + 5 \cdot 11 \cdot 13 &= 757
\end{aligned}
$$

Primes

$$
\begin{aligned}
2 \cdot 5 \cdot 7 + 3 \cdot 11 \cdot 13 &= 499 \\
3 \cdot 5 \cdot 7 + 2 \cdot 11 \cdot 13 &= 391 = 17 \cdot 23 \\
2 \cdot 3 \cdot 11 + 5 \cdot 7 \cdot 13 &= 521 \\
2 \cdot 5 \cdot 11 + 3 \cdot 7 \cdot 13 &= 383 \\
3 \cdot 5 \cdot 11 + 2 \cdot 7 \cdot 13 &= 347 \\
2 \cdot 7 \cdot 11 + 3 \cdot 5 \cdot 13 &= 349 \\
3 \cdot 7 \cdot 11 + 2 \cdot 5 \cdot 13 &= 361 = 19^2 \\
5 \cdot 7 \cdot 11 + 2 \cdot 3 \cdot 13 &= 463
\end{aligned}
$$

This time 11 out of 32 failed to be primes. It appears that the portion of primes is steadily decreasing. But will it ever reach 0? I doubt that will ever happen. Anyway, I came up with the following.

Problem. Let q_n be the number of primes in the set

$$\left\{ \prod_{p_j \in E} p_j + \prod_{p_k \in P_n \setminus E} p_k \,;\, E \subset P_n \right\}.$$

Then what is the greatest lower bound of

$$\frac{q_n}{2^{n-1}}?$$

Or rather, what is the value of

$$\liminf_{n \to \infty} \frac{q_n}{2^{n-1}}?$$

It would be a real shock to me if this limit turns out to be 0.

Chapter 23

Gaussian Integers

23.1 Gaussian Primes

Recall that to each point (x, y) in the plane, we can associate the complex number $z = x + iy$, where x and y are called the *real* and *imaginary parts* of z, respectively. The absolute value of a complex number $z = x + iy$ is, just like the case for real numbers, its distance to the origin:

$$|z| = \sqrt{x^2 + y^2}.$$

The conjugate of $z = x + iy$ is defined as $\bar{z} = x - iy$; it is the reflection of z in the real axis. We have

$$|z|^2 = z \cdot \bar{z}.$$

Exercise. Show that, for any two complex numbers z and w, we have

$$|zw| = |z| \cdot |w|;$$

i.e., the absolute value of the product of two complex numbers is the product of their absolute values.

Complex numbers of the form

$$a + bi, \text{ where } a \text{ and } b \text{ are (real) integers,}$$

are called *Gaussian integers*. Naturally, all real integers are Gaussian integers. Clearly, Gaussian integers correspond to lattice points in the plane. Furthermore, the sum and the product of two Gaussian integers are Gaussian integers.

Gaussian integers whose absolute values are 1, are called the *units*; there are four of them:

$$1, -1, i, -i.$$

Note that if z is a nonzero Gaussian integer, then $|z| \geq 1$.

Two Gaussian integers that differ by a unit factor are called *associates* or *equivalent*. For example, $1+i$ and $1-i$ are associates because $1+i = i(1-i)$. Note that each nonzero Gaussian integer has four associates (including itself), and they all have the same absolute value.

If Gaussian integer h is the product of two Gaussian integers f and g; i.e., if $h = fg$, then we say that f and g divide h; or f and g are the *divisors* of h. For example, $1+i$ is a divisor of 2 because $2 = (1+i)(1-i) = -i(1+i)^2$.

If a Gaussian integer h is the product of two or more divisors, and at least one of the divisors is neither a unit nor an associate of h, then this factorization of h is called *non-trivial*.

A Gaussian integer q is called *prime*, if it can only be factored in a trivial way. We saw that 2 is not a prime, nor is 5: $5 = (2+i)(2-i)$.

Exercise. Show that 3 and $1+i$ are Gaussian primes.

It is easy to see that if q is a Gaussian prime, then so is $\bar{q}$, and these two are distinct primes unless q is a real prime, or an associate of $1+i$.

Our purpose in this chapter is to describe all the primes among the Gaussian integers, but we need some preparation.

Two Gaussian integers h and k are called *relatively prime* if their only common divisors are units.

Lemma 1. For any pair of complex numbers h and q, where $q \neq 0$, there exists a Gaussian integer t such that

$$|h - tq| \leq \frac{|q|}{\sqrt{2}}.$$

Proof. This is obvious. Consider the set of lattice points

$$\{tq\,;\ t \text{ is a Gaussian integer}\}.$$

This set can be considered as the collection of the translates of four points

$$0, \quad q, \quad iq, \quad (1+i)q,$$

that are the vertices of a square (whose side length is $|q|$), and these squares cover the entire complex plane. (Figure GIP)

Gaussian Integers

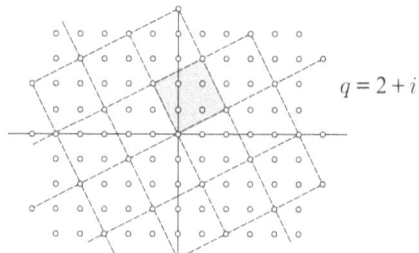

Figure GIP

Thus, for any complex number h, there must be a vertex of one of these squares that is at most $|q|/\sqrt{2}$ away from h; i.e.,

$$|h - t_0 q| \leq \frac{|q|}{\sqrt{2}} \quad \text{for some Gaussian integer } t_0.$$

□

Theorem 2. For an arbitrary pair of nonzero Gaussian integers h and k, there exists a pair of Gaussian integers r and s such that

$$rh + sk = q$$

is a greatest common divisor of h and k in the sense that q is a divisor of both h and k; and if t is a divisor of both h and k, then t is also a divisor of q.

Proof. If t is a divisor of both h and k, then for any pair of Gaussian integers r and s, clearly, t is a divisor of $q = rh + sk$. So we have only to prove the first part.

Of all the linear combinations $rh + sk$ of h and k (where r and s are Gaussian integers), let $q_0 = r_0 h + s_0 k$ be the one (one of the four) with the minimum non-zero absolute value. We claim that q_0 is a divisor of both h and k. By Lemma 1, for some Gaussian integer t_0, we have

$$|h - t_0 q_0| \leq \frac{|q_0|}{\sqrt{2}} < |q_0|.$$

But $h - t_0 q_0$ is also of the form $rh + sk$ (where $r = 1 - t_0 r_0$ and $s = -t_0 s_0$ are Gaussian integers), and q_0 is supposed to be the one with the minimum nonzero absolute value. Thus we must have

$$h - t_0 q_0 = 0; \quad \text{i.e.,} \quad h = t_0 q_0,$$

implying that q_0 is a divisor of h. Similarly, q_0 is also a divisor of k. □

Corollary 3. If h and k are relatively prime Gaussian integers, then 1 can be expressed as a linear combination of h and k: i.e.,

$$rh + sk = 1 \quad \text{for some Gaussian integers } r \text{ and } s.$$

Corollary 4. Let q be a Gaussian prime, and suppose that q divides the product $f \cdot g$ of two Gaussian integers f and g. Then q divides either f or g.

Proof. If q divides f, then we have nothing to prove. So, assume that q does not divide f. Then because q is a Gaussian prime, q and f are relatively prime. By Corollary 3, we have

$$rf + sq = 1 \quad \text{for some Gaussian integers } r \text{ and } s.$$

Multiplying this relation by g, we obtain

$$r(f \cdot g) + (sq)g = g.$$

By assumption, both terms on the left are divisible by q. Therefore, so is the right-hand side; i.e., g is divisible by q. □

We now proceed to describe all the Gaussian primes in a sequence of lemmas.

Lemma 5. If q is a Gaussian prime that is not a real integer (or an associate of a real integer), then $q\bar{q} = |q|^2$ is a real prime.

Proof. Set $c = |q|^2$. If c is not a real prime, then $c = ab$ for some real integers a and b that are not units. Hence $q\bar{q} = ab$; i.e., q divides the product of a and b. But q is a prime, so q must divide a or b. Say, q divides a; i.e., $a = qf$ for some Gaussian integer f. Note that f cannot be a unit, because the product of a unit and q that is not an associate of a real integer cannot be a real integer. We have

$$q\bar{q} = (qf) \cdot b, \quad \text{and so } \bar{q} = f \cdot b.$$

This is a non-trivial factorization of $\bar{q}$ that is a Gaussian prime. We obtain a contradiction. Hence $|q|^2$ must be a real prime. □

Conversely, we have the following.

Lemma 6. A real prime p that is not a Gaussian prime must be of the form $p = q \cdot \bar{q} = |q|^2$ for some Gaussian prime q. Furthermore, this expression is essentially unique.

Proof. For a real prime p that is not a Gaussian prime there must be a non-trivial factorization of p; i.e., $p = q \cdot f$, where q is a Gaussian prime that is not an associate of a real integer, and f is not a unit. Taking complex conjugate of both sides, we

Gaussian Integers

obtain $p = \bar{q} \cdot \bar{f}$; i.e., $q \cdot f = \bar{q} \cdot \bar{f}$. It follows that q divides the product of $\bar{q}$ and $\bar{f}$. But q is a Gaussian prime, so q must divide either $\bar{q}$ or $\bar{f}$. Suppose Gaussian prime q divides its complex conjugate $\bar{q}$. This can happen if and only if $q = 1+i$ or an associate of $1+i$. (Assume that $\bar{q} = qw$. Then because $|\bar{q}| = |q|$, we must have $|w| = 1$; i.e., w is a unit. But $w = \pm 1$ implies that q is either real or an associate of a real. And $w = \pm i$ implies that q is an associate of $1+i$.) Without loss of generality, we may assume that $q = 1+i$. We have $p = (1+i) \cdot f$. Considering the arguments of both sides, we obtain

$$\arg(1+i) + \arg f \equiv 0, \quad \arg f \equiv -\frac{\pi}{4} \pmod{2\pi}.$$

Hence $f = (1-i)u$ for some real integer u. Substituting this result into $p = q \cdot f$, we obtain

$$p = (1+i)(1-i)u = 2u,$$

and so p is an even prime; i.e., $p = 2$ (and $u = 1$).

On the other hand, if q divides $\bar{f}$, then $\bar{f} = q \cdot f_0$ for some Gaussian integer f_0. And we have

$$p = \bar{q} \cdot \bar{f} = \bar{q} \cdot (q \cdot f_0) = (q\bar{q})f_0 = |q|^2 \cdot f_0.$$

Thus a real integer $|q|^2$ divides p, but p is a real prime, so we must have $p = |q|^2$ (and $f_0 = 1$).

Now, by our assumption, q can be expressed as $q = a + bi$, where a and b are nonzero real integers; and so

$$p = |q|^2 = a^2 + b^2.$$

We obtain: A real prime that is not a Gaussian prime can be expressed as the sum of two perfect squares.

Note that this expression is essentially unique. Because if there exists another pair of nonzero real integers a_0 and b_0 such that

$$p = a^2 + b^2 = a_0^2 + b_0^2; \quad \text{i.e., } p = q \cdot \bar{q} = q_0 \cdot \bar{q}_0,$$

where $q_0 = a_0 + b_0 i$, then q divides the product of q_0 and $\bar{q}_0$. But q is a Gaussian prime, so q must divide either q_0 or $\bar{q}_0$. We may assume, without loss of generality, that q divides q_0. (If necessary, we may interchange the notations q_0 and $\bar{q}_0$; i.e., $a_0 + b_0 i$ and $a_0 - b_0 i$.) Thus

$$q_0 = q \cdot f_0 \quad \text{for some Gaussian integer } f_0.$$

But then

$$q \cdot \bar{q} = q_0 \cdot \bar{q}_0 = (q \cdot f_0) \cdot \bar{q}_0.$$

Dividing both sides by q, we obtain $\bar{q} = f_0 \cdot \bar{q}_0$. Now $\bar{q}$ is a Gaussian prime, so this factorization of $\bar{q}$ must be trivial. Hence the expression

$$p = a^2 + b^2$$

is unique other than the order of the summands. □

Combining these two lemmas, and noting that the sum of two perfect squares cannot be congruent to -1 (mod 4), we obtain the following.

Theorem 7. A real prime p is not a Gaussian prime if and only if p can be expressed as the sum of two (nonzero) perfect squares; i.e., $p = a^2 + b^2$ for some positive integers a and b.
And so the (real) primes of the form $4n - 1$ must be Gaussian primes.

Note carefully, we still do not know (real) primes of the form $4n+1$ are Gaussian primes or not (because we don't know every such prime can be expressed as the sum of two perfect squares). Next two lemmas settle this problem.

Lemma 8. For a (real) prime p of the form $4n + 1$, equation $x^2 + 1 \equiv 0$ (mod p) has a (real) integer solution.

Proof. This lemma is an immediate consequence of the Wilson theorem, which asserts that, for any (real) prime p,

$$(p-1)! + 1 \equiv 0 \pmod{p}.$$

For example, if $p = 13$ say, then we have

$$\begin{aligned} 0 &\equiv 12! + 1 \equiv 6! \cdot (7 \cdot 8 \cdots 11 \cdot 12) + 1 \\ &\equiv (6!) \cdot (-6) \cdot (-5) \cdots (-2) \cdot (-1) + 1 \\ &\equiv (-1)^6 \cdot (6!)^2 + 1 \pmod{13} \end{aligned}$$

Hence $x^2 + 1 \equiv 0$ (mod 13) has an integer solution $x \equiv 6!$ (mod 13). It is easy to see that this reasoning can be extended to any prime of the form $4n + 1$. However, instead of presenting a proof of the Wilson Theorem, we give a direct proof of the lemma by Roger Heath-Brown.[1]

We start with some preparation. For any integer a that is not a multiple of p, we claim that no two of the numbers

$$a, 2a, \cdots, (p-1)a$$

are congruent to each other (mod p). For, if $1 \leq r < s \leq p - 1$, then

$$sa - ra = (s - r)a \not\equiv 0 \pmod{p}.$$

Therefore, the residues (mod p) of $p - 1$ numbers

$$a, 2a, \cdots, (p-1)a,$$

---
[1] *Fermat's two square theorem*, Invariant (1984), 2-5.

Gaussian Integers

are all distinct, and so by the pigeon-hole principle, they are precisely

$$1, 2, \cdots, (p-1), \quad \text{in some order}.$$

In particular, one of them, say $\tilde{a}a$, must satisfy

$$\tilde{a}a \equiv 1 \pmod{p}.$$

This $\tilde{a}$ is called the *multiplicative inverse* of a (mod p). Clearly, each of 1 and $(p-1)$ are multiplicative inverse of itself. But none of the others is a multiplicative inverse of itself because

$$r^2 - 1 = (r-1)(r+1) \equiv 0 \pmod{p}$$

implies that $r \equiv \pm 1 \pmod{p}$.

Now we partition the set of residues $\{1, 2, \cdots, p-1\}$ into classes such that each element and its additive and multiplicative inverses are in the same class:

$$\{x, -x, \tilde{x}, -\tilde{x}\}$$

(where $-x \equiv p - x$ and $x\tilde{x} \equiv 1 \pmod{p}$). We know for $x = 1$, this class reduces to a class of (only) 2 elements; i.e., $\{1, p-1\}$. But our set of residues $\{1, 2, \cdots, p-1\}$ has $p - 1 = 4n$ elements. So there must be another class that has only 2 elements. But $x \equiv -x \pmod{p}$ cannot occur for an odd prime p, while $x \equiv \tilde{x} \pmod{p}$ implies $x^2 \equiv x\tilde{x} \equiv 1 \pmod{p}$, and this can occur only for $x \equiv 1$ or $x \equiv p - 1 \pmod{p}$. So for the other class to reduce to a 2-element set, we must have $x \equiv -\tilde{x} \pmod{p}$, which implies that

$$x^2 \equiv -x\tilde{x} \equiv -1 \pmod{p}.$$

We have shown that for $p = 4n+1$, equation $x^2 + 1 \equiv 0 \pmod{p}$ has a solution. $\square$

Lemma 9. *For equation $x^2 + 1 \equiv 0 \pmod{p}$, where p is a (real) prime, to have a (real) integer solution, p cannot be a Gaussian prime.*

Proof. Note that, as a Gaussian integer, $x^2 + 1$ can be factored. In fact, we have

$$(x+i) \cdot (x-i) \equiv 0 \pmod{p}.$$

So if p is a Gaussian prime, then p must divide one of the factors on the left; i.e.,

$$x \pm i = p(a + bi)$$

for some real integers a and b. But then $pb = \pm 1$, which is absurd. $\square$

Note that, combining the last two lemmas, we have shown that a real prime of the form $4n + 1$ cannot be a Gaussian prime.

Corollary 10. If an odd (real) integer k divides an integer of the form $x^2 + 1$ (where x is a real integer), then k must be of the form $4n + 1$.

Corollary 11. A (real) prime p of the form $4n + 1$ can be expressed uniquely as the sum of two (nonzero) perfect squares.

Summing up, we obtain the following:

Theorem 12.

(a) The integer 2 is not a Gaussian prime.

(b) Every real prime p of the form $4n - 1$ is a Gaussian prime.

(c) Every real prime p of the form $4n + 1$ can be factored, essentially uniquely, as $p = q \cdot \bar{q}$, where q is a Gaussian prime.

(d) All the Gaussian primes are accounted for in parts a, b, and c.

Exercise.

(a) Show that the set of all positive integers that can be expressed as the sums of at most two perfect squares is closed under multiplication; i.e., the product of two such integers can also be expressed as the sum of at most two perfect squares. (*Hint*: Exercise on page 209.)

(b) Describe all the (real) integers that can be expressed as the sum of at most two perfect squares.

23.2 An Application to Real Primes

As a by-product of our discussion, we obtain the following.

Theorem 14. There exist infinitely many (real) primes of the form $4n+1$.

Proof. Suppose there exist only finitely many such primes, say $q_1, q_2, \cdots, q_k$. Set

$$m = (2 \cdot q_1 \cdot q_2 \cdots q_k)^2 + 1.$$

Note that m is odd, and none of the primes $2, q_1, q_2, \cdots, q_k$ divides m. So m has a (real) prime factor (possibly m itself) that is not any of $2, q_1, q_2, \cdots, q_k$. By Corollary 10 above, we have shown that there exists a new prime of the form $4n + 1$. □

Recall that any (real) nonzero integer can be expressed uniquely as the product of (positive) prime numbers and a power of -1. (Naturally, the exponent of -1 is considered mod 2 for uniqueness.) Gaussian integers also satisfy the unique factorization theorem; i.e., any nonzero Gaussian integer can be expressed uniquely as the product of "positive" Gaussian primes and a power of i, where the exponent of i is considered

Gaussian Integers

mod 4, provided we agree to call a Gaussian integer z "*positive*" if $-\frac{\pi}{4} < \arg z \leq \frac{\pi}{4}$. Note that every (nonzero) Gaussian integer has one and only one associate that is "positive" in this sense. The proof of the unique factorization theorem is similar to the real case and so we leave it as an exercise for the reader.

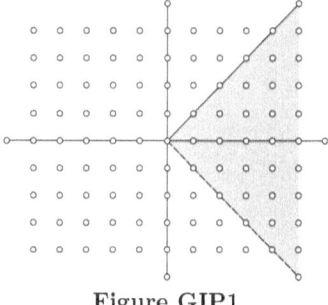

Figure GIP1

Exercise. Is the set of all "*positive*" Gaussian integers closed under addition and multiplication?

Chapter 24

Calculus with Complex Numbers

Consider the differential equation

$$f'(x) = f(x).$$

Clearly, $f(x) = e^x$ is a solution, so is $f(x) = ce^x$ for any constant c. To determine the constant c, we need an initial condition such as $f(0) = 3$. Then the solution becomes unique: $f(x) = 3e^x$. More generally, the solution of the differential equation

$$f'(x) = kf(x)$$

must be the form $f(x) = ce^{kx}$, for some constant c. And we need an initial condition to determine the solution uniquely.

Now consider the differential equation

$$f''(x) = 4f(x).$$

Clearly, $f(x) = ce^{2x}$ is a solution, so is $f(x) = ce^{-2x}$ for any constant c. In fact, its general solution must be of the form

$$f(x) = c_1 e^{2x} + c_2 e^{-2x}$$

for some pair of constants c_1 and c_2. For the solution to be unique, we need initial conditions such as $f(0) = 1$, $f'(0) = -1$. Then

$$f(0) = c_1 + c_2 = 1, \quad f'(0) = 2c_1 - 2c_2 = -1.$$

Therefore,

$$c_1 = \frac{1}{4}, \quad c_2 = \frac{3}{4}; \quad f(x) = \frac{1}{4}\left(x^{2x} + 3e^{-2x}\right).$$

For the differential equation
$$f''(x) = kf(x),$$
the general solution is
$$f(x) = c_1 e^{\sqrt{k}x} + c_2 e^{-\sqrt{k}x},$$
where the constants c_1 and c_2 are determined by the initial condition.

Now if the reader has no objection about our discussion so far, then consider the case $k = -1$; i.e., consider the differential equation
$$f''(x) = -f(x).$$
Its general solution must be of the form
$$f(x) = c_1 e^{ix} + c_2 e^{-ix} \quad (i = \sqrt{-1}).$$
However, it is easy to see that both
$$f(x) = \cos x \quad \text{and} \quad f(x) = \sin x$$
satisfy this differential equation. What are the initial conditions for these two functions? Certainly, for $f(x) = \cos x$, we have
$$f(0) = [\cos x]_{x=0} = 1 \quad \text{and} \quad f'(0) = [-\sin x]_{x=0} = 0.$$
Hence $c_1 + c_2 = 1$, $i(c_1 - c_2) = 0$; i.e., $c_1 = c_2 = \frac{1}{2}$. We obtain
$$\cos x = \frac{e^{ix} + e^{-ix}}{2}.$$
Similarly, for $f(x) = \sin x$, we have
$$f(0) = [\sin x]_{x=0} = 0 \quad \text{and} \quad f'(0) = [\cos x]_{x=0} = 1.$$
Hence $c_1 + c_2 = 0$, $i(c_1 - c_2) = 1$; i.e., $c_1 = -c_2 = \frac{1}{2i}$. We obtain
$$\sin x = \frac{e^{ix} - e^{-ix}}{2i}.$$
Note that these identities express trigonometric functions in terms of the exponential function. Now these two identities give the so-called *Euler identity*
$$e^{ix} = \cos x + i \sin x, \quad e^{-ix} = \cos x - i \sin x.$$

(Note that, by replacing x by $-x$, or by taking the complex conjugate of both sides, we see that these two identities are actually equivalent.) This is an amazing identity. In particular, the Euler identity says that the exponential function must be periodic. (What is its period?)

Calculus with Complex Numbers

Setting $x = \pi$ in the Euler identity, we obtain

$$e^{i\pi} = \cos \pi + i \sin \pi = -1.$$

Rewriting, we obtain

$$e^{i\pi} + 1 = 0.$$

This is a remarkable equality. It combines the five most important constants 0, 1, π, e, i together. (What if we set $x = -\pi$?)

Many identities in trigonometry follow from the Euler identity

$$e^{ix} = \cos x + i \sin x.$$

For example, multiply the Euler identity with its complex conjugate; i.e., multiplying the equalities

$$e^{ix} = \cos x + i \sin x \quad \text{and} \quad e^{-ix} = \cos x - i \sin x,$$

we obtain

$$e^{ix} \cdot e^{-ix} = (\cos x)^2 - (i \sin x)^2;$$

which is the identity

$$1 = \cos^2 x + \sin^2 x.$$

Actually, this is a particular case of the following. Multiply

$$e^{ix} = \cos x + i \sin x \quad \text{and} \quad e^{iy} = \cos y + i \sin y,$$

we obtain

$$e^{ix} \cdot e^{iy} = (\cos x \cdot \cos y - \sin x \cdot \sin y) + i(\sin x \cdot \cos y + \cos x \cdot \sin y).$$

But

$$e^{ix} \cdot e^{iy} = e^{i(x+y)} = \cos(x + y) + i \sin(x + y);$$

so by comparing the real and imaginary parts of these two expressions, we obtain the addition formula for the trigonometric functions

$$\begin{aligned} \cos(x + y) &= \cos x \cdot \cos y - \sin x \cdot \sin y, \\ \sin(x + y) &= \sin x \cdot \cos y + \cos x \cdot \sin y. \end{aligned}$$

Or, cube both sides of the Euler identity

$$e^{ix} = \cos x + i \sin x,$$

we obtain

$$\begin{aligned} e^{3ix} &= \left(e^{ix}\right)^3 = (\cos x + i \sin x)^3 \\ &= \cos^3 x + 3 \cos^2 x \cdot (i \sin x) + 3 \cos x \cdot (i \sin x)^2 + (i \sin x)^3. \end{aligned}$$

Comparing the real and imaginary parts of both sides, we obtain

$$\cos 3x = \cos^3 x - 3\cos x \cdot \sin^2 x = 4\cos^3 x - 3\cos x,$$
$$\sin 3x = 3\cos^2 x \cdot \sin x - \sin^3 x = 3\sin x - 4\sin^3 x.$$

One final example (where $\Im$ is a short-hand for "the imaginary part of").

$$\sin x + \sin 2x + \sin 3x + \cdots + \sin nx$$
$$= \Im\left\{e^{ix} + e^{2ix} + e^{3ix} + \cdots + e^{inx}\right\}$$
$$= \Im\left\{\frac{e^{ix} - e^{i(n+1)x}}{1 - e^{ix}}\right\} = \Im\left\{\frac{e^{i(n+\frac{1}{2})x} - e^{i\frac{x}{2}}}{e^{i\frac{x}{2}} - e^{-i\frac{x}{2}}}\right\}$$
$$= \Im\frac{\{\cos(n+\frac{1}{2})x - \cos\frac{x}{2}\} - i\{\cdots\}}{2i\sin\frac{x}{2}}$$
$$= \frac{\cos\frac{x}{2} - \cos(n+\frac{1}{2})x}{2\sin\frac{x}{2}} = \frac{\sin\frac{n}{2}x \cdot \sin\frac{n+1}{2}x}{\sin\frac{x}{2}} \quad (x \not\equiv 0 \pmod{2\pi}).$$

(What if we take the real part? See also page 153.)

Dividing the two extreme sides by x, and taking the limit as $x \to 0$ (and noting that $\lim_{n \to 0} \frac{\sin kx}{x} = k$), we obtain

$$1 + 2 + 3 + \cdots + n = \lim_{x \to 0}\left\{\left(\frac{\sin\frac{n}{2}x}{x}\right) \cdot \left(\frac{\sin\frac{n+1}{2}x}{x}\right) \cdot \left(\frac{x}{\sin\frac{x}{2}}\right)\right\}$$
$$= \frac{n}{2} \cdot \frac{n+1}{2} \cdot \frac{1}{\frac{1}{2}} = \frac{n(n+1)}{2}.$$

(Note that this result helps us remember the formula above.)

What is the moral? Mathematics becomes much more transparent, fruitful and easy if we work in the complex field. Life is much harder and complicated if we confine ourselves to the realm of the real numbers. Take the fundamental theorem of algebra for example. It asserts that: Every polynomial of degree n has precisely n roots. This simple yet very useful statement is true only if we are working in the complex field (but not in the real field).

In conclusion, we quote J. Hadamard: "The shortest path between two truths in the real domain passes through the complex domain."

Readers who want to know more about the use of complex numbers in calculus will find much to enjoy in *Classical Complex Analysis*, co-authored with Bernard Epstein (Jones and Bartlett Publishers, Sudbury, MA; 1996).

Appendix

Determinants

A.1 Genesis

The origin of determinants lies in the investigation of the general solutions of system of simultaneous linear equations. Historically, determinants first appeared in works (1683) of SEKI Takakazu (Kōwa) of Japan ahead of the letters (1693) of Leibnitz to l'Hôpital. In 1750, Cramer attracted attention by exhibiting solutions of systems of simultaneous linear equations in arbitrarily many unknowns, and thus stimulated Cauchy and Jacobi to develop the theory of determinants.

Rather than defining determinants "out of blue", we shall present them as a natural consequence of the study of solutions of systems of simultaneous linear equations as was originally developed by Seki, Leibnitz and Cramer.

It is well-known that in solving a given system of simultaneous linear equations with two unknowns,

$$\begin{cases} a_1 x + b_1 y = k_1, \\ a_2 x + b_2 y = k_2, \end{cases}$$

we eliminate one unknown to obtain an equation in the remaining unknown. This is done by multiplying each of the equations by b_2 and $-b_1$, or by $-a_2$ and a_1, then adding the resulting equations. In this way we obtain

$$\begin{aligned} (a_1 b_2 - a_2 b_1) x &= k_1 b_2 - k_2 b_1, \\ (a_1 b_2 - a_2 b_1) y &= a_1 k_2 - a_2 k_1. \end{aligned}$$

Therefore (assuming $a_1 b_2 - a_2 b_1 \neq 0$), we obtain

$$x = \frac{k_1 b_2 - k_2 b_1}{a_1 b_2 - a_2 b_1}, \quad y = \frac{a_1 k_2 - a_2 k_1}{a_1 b_2 - a_2 b_1}.$$

It is clear that such a method is applicable even in cases involving more than two unknowns. However, it is not simple to anticipate the final expressions. Yet if we take a close look at the formulas obtained above, we can discover a clue to its generalization.

We notice that both the numerator and the denominator have the same form, $u_1v_2 - u_2v_1$. This expression has the properties that (i) it is linear and homogenous in u_1, u_2, and in v_1, v_2; i.e., in $\begin{bmatrix} u_1 \\ u_2 \end{bmatrix}$ and in $\begin{bmatrix} v_1 \\ v_2 \end{bmatrix}$. Moreover, (ii) if these two vectors coincide; i.e., $\begin{bmatrix} u_1 \\ u_2 \end{bmatrix} = \begin{bmatrix} v_1 \\ v_2 \end{bmatrix}$, then the expression vanishes.

With these two as the only postulates, we can derive formulas for solutions of systems of simultaneous linear equations, but first let us explain our terminology.

Definition 1. A function $\Delta(\vec{a}_1, \vec{a}_2, \cdots, \vec{a}_n)$ of n (column) vectors $\vec{a}_1$, $\vec{a}_2, \cdots, \vec{a}_n$ in the n-space is said to be *multilinear* if it is linear with respect to each vector $\vec{a}_i$ ($1 \leq i \leq n$) when the other vectors are kept fixed; namely,

$$\Delta(\vec{a}_1, \vec{a}_2, \cdots, \vec{a}_{i-1}, t\vec{a}_i + s\vec{a}'_i, \vec{a}_{i+1}, \cdots, \vec{a}_n)$$
$$= t \cdot \Delta(\vec{a}_1, \cdots, \vec{a}_i, \cdots, \vec{v}_n) + s \cdot \Delta(\vec{a}_1, \cdots, \vec{a}'_i, \cdots, \vec{v}_n).$$

For example, for $\vec{a}_1 = \begin{bmatrix} u_1 \\ u_2 \end{bmatrix}$, $\vec{a}_2 = \begin{bmatrix} v_1 \\ v_2 \end{bmatrix}$, if we define

$$\Delta(\vec{a}_1, \vec{a}_2) = \Delta \begin{bmatrix} u_1 & v_1 \\ u_2 & v_2 \end{bmatrix} = u_1v_2 - u_2v_1,$$

then, with obvious notations, we have

$$\Delta(t\vec{a}_1 + s\vec{a}'_1, \vec{a}_2) = \Delta \begin{bmatrix} tu_1 + su'_1 & v_1 \\ tu_2 + su'_2 & v_2 \end{bmatrix}$$
$$= (tu_1 + su'_1)v_2 - (tu_2 + su'_2)v_1$$
$$= t(u_1v_2 - u_2v_1) + s(u'_1v_2 - u'_2v_1)$$
$$= t \cdot \Delta \begin{bmatrix} u_1 & v_1 \\ u_2 & v_2 \end{bmatrix} + s \cdot \Delta \begin{bmatrix} u'_1 & v_1 \\ u'_2 & v_2 \end{bmatrix}$$
$$= t \cdot \Delta(\vec{a}_1, \vec{a}_2) + s \cdot \Delta(\vec{a}'_1, \vec{a}_2).$$

The same holds for the second (column) vector $\vec{a}_2$.

Clearly, a multilinear function of vectors $\vec{a}_1, \cdots, \vec{a}_n$ vanishes whenever one of the vectors is the zero vector.

Definition 2. A function $\Delta(\vec{a}_1, \vec{a}_2, \cdots, \vec{a}_n)$ of n (column) vectors $\vec{a}_1$, $\vec{a}_2, \cdots, \vec{a}_n$ in the n-space is said to be *alternate* if it vanishes whenever any two of the vectors $\vec{a}_1, \vec{a}_2, \cdots, \vec{a}_n$ coincide.

Clearly, $\Delta(\vec{a}_1, \vec{a}_2)$ in the example above is alternate.

Determinants

These concepts are all very simple, but the following reasoning depends only on the multilinearity and "alternateness" of the function Δ. Set

$$f_1 \equiv a_1 x + b_1 y - k_1 = 0, \qquad f_2 \equiv a_2 x + b_2 y - k_2 = 0,$$

$$\vec{\alpha} = \begin{bmatrix} a_1 \\ a_2 \end{bmatrix}, \quad \vec{\beta} = \begin{bmatrix} b_1 \\ b_2 \end{bmatrix}, \quad \vec{\gamma} = \begin{bmatrix} k_1 \\ k_2 \end{bmatrix}, \quad \vec{\delta} = \begin{bmatrix} f_1 \\ f_2 \end{bmatrix}.$$

Then $\vec{\delta} = x\vec{\alpha} + y\vec{\beta} - \vec{\gamma}$, and so for any alternate multilinear function Δ of two vectors, we have

$$\begin{aligned} \Delta(\vec{\delta}, \vec{\beta}) &= \Delta(x\vec{\alpha} + y\vec{\beta} - \vec{\gamma}, \vec{\beta}) \\ &= x \cdot \Delta(\vec{\alpha}, \vec{\beta}) + y \cdot \Delta(\vec{\beta}, \vec{\beta}) - \Delta(\vec{\gamma}, \vec{\beta}) \\ &= x \cdot \Delta(\vec{\alpha}, \vec{\beta}) - \Delta(\vec{\gamma}, \vec{\beta}). \end{aligned}$$

Hence, if x, y satisfy the system of simultaneous equations, then $\vec{\delta} = \begin{bmatrix} f_1 \\ f_2 \end{bmatrix} = \begin{bmatrix} 0 \\ 0 \end{bmatrix} = \vec{0}$, and so $\Delta(\vec{\delta}, \vec{\beta}) = \Delta(\vec{0}, \vec{\beta}) = 0$. It follows that (assuming $\Delta(\vec{\alpha}, \vec{\beta}) \neq 0$)

$$x \cdot \Delta(\vec{\alpha}, \vec{\beta}) - \Delta(\vec{\gamma}, \vec{\beta}) = 0. \qquad x = \frac{\Delta(\vec{\gamma}, \vec{\beta})}{\Delta(\vec{\alpha}, \vec{\beta})}.$$

This is in agreement with our preceding result. Similarly, we can also solve for y.

Our reasoning is valid with an arbitrary number of unknowns. Consider a system of simultaneous linear equations with n unknowns $x_1, x_2, \cdots, x_n$:

$$\begin{cases} f_1 \equiv a_{11}x_1 + a_{12}x_2 + \cdots + a_{1n}x_n - k_n = 0, \\ f_2 \equiv a_{21}x_1 + a_{22}x_2 + \cdots + a_{2n}x_n - k_n = 0, \\ \cdots \\ f_n \equiv a_{n1}x_1 + a_{n2}x_2 + \cdots + a_{nn}x_n - k_n = 0. \end{cases}$$

Let us assume that there exists a function Δ of n (column) vectors $\vec{\alpha}_1, \vec{\alpha}_2, \cdots, \vec{\alpha}_n$, in the n-space with the following two properties:

(a) Δ is multilinear;

(b) Δ is alternate.

We write

$$\Delta(\vec{\alpha}_1, \vec{\alpha}_2, \cdots, \vec{\alpha}_n) = \Delta \begin{bmatrix} a_{11} & a_{12} & \cdots & a_{1n} \\ a_{21} & a_{22} & \cdots & a_{2n} \\ \cdots & \cdots & & \cdots \\ a_{n1} & a_{n2} & \cdots & a_{nn} \end{bmatrix} = \Delta(A),$$

where $\vec{\alpha}_j = \begin{bmatrix} a_{1j} \\ a_{2j} \\ \vdots \\ a_{nj} \end{bmatrix}$. Because we are assuming the existence of an alternate multilinear function, we can eliminate all but one of the unknowns as before: To solve for x_i $(1 \le i \le n)$, we replace the ith vector $\vec{\alpha}_i$ in Δ by $\vec{\delta} = \begin{bmatrix} f_1 \\ f_2 \\ \vdots \\ f_n \end{bmatrix} = \sum_{j=1}^n x_j \vec{\alpha}_j - \vec{\gamma}$,

where $\vec{\gamma} = \begin{bmatrix} k_1 \\ k_2 \\ \vdots \\ k_n \end{bmatrix}$. Then

$$\Delta\left(\vec{\alpha}_1, \cdots, \vec{\alpha}_{i-1}, \vec{\delta}, \vec{\alpha}_{i+1}, \cdots, \vec{\alpha}_n\right)$$
$$= \Delta\left(\vec{\alpha}_1, \cdots, \vec{\alpha}_{i-1}, \sum_{j=1}^n x_j \vec{\alpha}_j - \vec{\gamma}, \vec{\alpha}_{i+1}, \cdots, \vec{\alpha}_n\right)$$
$$= \sum_{j=1}^n x_j \cdot \Delta\left(\vec{\alpha}_1, \cdots, \vec{\alpha}_{i-1}, \vec{\alpha}_j, \vec{\alpha}_{i+1}, \cdots, \vec{\alpha}_n\right)$$
$$\quad - \Delta\left(\vec{\alpha}_1, \cdots, \vec{\alpha}_{i-1}, \vec{\gamma}, \vec{\alpha}_{i+1}, \cdots, \vec{\alpha}_n\right) \quad \text{(by multilinearity)}$$
$$= x_i \cdot \Delta\left(\vec{\alpha}_1, \cdots, \vec{\alpha}_{i-1}, \vec{\alpha}_i, \vec{\alpha}_{i+1}, \cdots, \vec{\alpha}_n\right)$$
$$\quad - \Delta\left(\vec{\alpha}_1, \cdots, \vec{\alpha}_{i-1}, \vec{\gamma}, \vec{\alpha}_{i+1}, \cdots, \vec{\alpha}_n\right) \quad \text{(by alternateness)}.$$

If $\begin{bmatrix} x_1 \\ x_2 \\ \vdots \\ x_n \end{bmatrix}$ is a solution to the system of simultaneous linear equations, then all f_i are 0, and so $\vec{\delta} = \vec{0}$. Hence we obtain

$$x_i \cdot \Delta\left(\vec{\alpha}_1, \cdots, \vec{\alpha}_{i-1}, \vec{\alpha}_i, \vec{\alpha}_{i+1}, \cdots, \vec{\alpha}_n\right)$$
$$- \Delta\left(\vec{\alpha}_1, \cdots, \vec{\alpha}_{i-1}, \vec{\gamma}, \vec{\alpha}_{i+1}, \cdots, \vec{\alpha}_n\right) = 0;$$

i.e., (assuming the denominator does not vanish)

$$x_i = \frac{\Delta\left(\vec{\alpha}_1, \cdots, \vec{\alpha}_{i-1}, \vec{\gamma}, \vec{\alpha}_{i+1}, \cdots, \vec{\alpha}_n\right)}{\Delta\left(\vec{\alpha}_1, \cdots, \vec{\alpha}_{i-1}, \vec{\alpha}_i, \vec{\alpha}_{i+1}, \cdots, \vec{\alpha}_n\right)} \quad (1 \le i \le n).$$

The result above was obtained by *assuming* the existence of an alternate multilinear function of n vectors in the n-space. We now want to *prove* the existence of such a function and discuss its properties, but let us start from the following.

Determinants

Lemma 3. Suppose $\Delta(\vec{\alpha}_1, \vec{\alpha}_2, \cdots, \vec{\alpha}_n)$ is a multilinear function of n vectors in the n-space. Then the following are equivalent:

(a) Δ is alternate; i.e., $\Delta(\vec{\alpha}_1, \vec{\alpha}_2, \cdots, \vec{\alpha}_n) = 0$ whenever two vectors are equal.

(b) Δ changes sign whenever a pair of vectors are interchanged; i.e.,

$$\Delta(\vec{\alpha}_{\sigma 1}, \vec{\alpha}_{\sigma 2}, \cdots, \vec{\alpha}_{\sigma n}) = -\Delta(\vec{\alpha}_1, \vec{\alpha}_2, \cdots, \vec{\alpha}_n),$$

where $\sigma = (i, j)$ interchanges the indices i and j, $i \neq j$ ($1 \leq i \leq n$, $1 \leq j \leq n$); i.e.,

$$\sigma k = \begin{cases} k & (k \neq i, \, k \neq j), \\ i & (k = j), \\ j & (k = i). \end{cases}$$

Proof. (a) $\Longrightarrow$ (b). Suppose $\Delta(\vec{\alpha}_1, \vec{\alpha}_2, \cdots, \vec{\alpha}_n)$ is an alternate multilinear function of n vectors $\vec{\alpha}_1, \vec{\alpha}_2, \cdots, \vec{\alpha}_n$, and $1 \leq i < j \leq n$. Set

$$\Delta_0(\vec{\alpha}_i, \vec{\alpha}_j) = \Delta(\vec{\alpha}_1, \vec{\alpha}_2, \cdots, \vec{\alpha}_n);$$

i.e., consider Δ_0 as a function of two vectors $\vec{\alpha}_i$ and $\vec{\alpha}_j$ only (all other vectors are kept fixed). Then by the multilinearity, we have

$$\begin{aligned}\Delta_0(\vec{\alpha}_i + \vec{\alpha}_j, \vec{\alpha}_i + \vec{\alpha}_j) &= \Delta_0(\vec{\alpha}_i, \vec{\alpha}_i) + \Delta_0(\vec{\alpha}_i, \vec{\alpha}_j) \\ &\quad + \Delta_0(\vec{\alpha}_j, \vec{\alpha}_i) + \Delta_0(\vec{\alpha}_j, \vec{\alpha}_j).\end{aligned}$$

But Δ_0 is alternate, so

$$\Delta_0(\vec{\alpha}_i + \vec{\alpha}_j, \vec{\alpha}_i + \vec{\alpha}_j) = \Delta_0(\vec{\alpha}_i, \vec{\alpha}_i) = \Delta_0(\vec{\alpha}_j, \vec{\alpha}_j) = 0.$$

It follows that

$$\Delta_0(\vec{\alpha}_i, \vec{\alpha}_j) + \Delta_0(\vec{\alpha}_j, \vec{\alpha}_i) = 0.$$

(b) $\Longrightarrow$ (a). Conversely, suppose Δ changes sign when two vectors are interchanged. We have, using the notation above,

$$\Delta_0(\vec{\alpha}_i, \vec{\alpha}_j) = -\Delta_0(\vec{\alpha}_j, \vec{\alpha}_i).$$

If $\vec{\alpha}_i = \vec{\alpha}_j$, then this gives

$$(1+1) \cdot \Delta_0(\vec{\alpha}_i, \vec{\alpha}_i) = 0, \quad \Delta_0(\vec{\alpha}_i, \vec{\alpha}_i) = 0.$$

$\square$

Example 4. Let $\vec{e}_1 = \begin{bmatrix} 1 \\ 0 \end{bmatrix}$, $\vec{e}_2 = \begin{bmatrix} 0 \\ 1 \end{bmatrix}$, and $A = \begin{bmatrix} a_1 & b_1 \\ a_2 & b_2 \end{bmatrix}$. Then

$$\vec{a} = \begin{bmatrix} a_1 \\ a_2 \end{bmatrix} = a_1 \vec{e}_1 + a_2 \vec{e}_2, \quad \vec{b} = \begin{bmatrix} b_1 \\ b_2 \end{bmatrix} = b_1 \vec{e}_1 + b_2 \vec{e}_2.$$

Hence if $\Delta(A) = \Delta(\vec{a}, \vec{b}) = \Delta\begin{bmatrix} a_1 & b_1 \\ a_2 & b_2 \end{bmatrix}$ is an alternate multilinear function of (column) vectors, then

$$\begin{aligned}
\Delta(A) &= \Delta(\vec{a}, \vec{b}) = \Delta\begin{bmatrix} a_1 & b_1 \\ a_2 & b_2 \end{bmatrix} \\
&= \Delta\left(a_1\vec{e}_1 + a_2\vec{e}_2,\ b_1\vec{e}_1 + b_2\vec{e}_2\right) \\
&= \Delta\left(a_1\vec{e}_1,\ b_1\vec{e}_1 + b_2\vec{e}_2\right) + \Delta\left(a_2\vec{e}_2,\ b_1\vec{e}_1 + b_2\vec{e}_2\right) \\
&= \Delta\left(a_1\vec{e}_1,\ b_1\vec{e}_1\right) + \Delta\left(a_1\vec{e}_1,\ b_2\vec{e}_2\right) \\
&\quad + \Delta\left(a_2\vec{e}_2,\ b_1\vec{e}_1\right) + \Delta\left(a_2\vec{e}_2,\ b_2\vec{e}_2\right) \\
&= a_1 b_1 \cdot \Delta\left(\vec{e}_1, \vec{e}_1\right) + a_1 b_2 \cdot \Delta\left(\vec{e}_1, \vec{e}_2\right) \\
&\quad + a_2 b_1 \cdot \Delta\left(\vec{e}_2, \vec{e}_1\right) + a_2 b_2 \cdot \Delta\left(\vec{e}_2, \vec{e}_2\right) \\
&= (a_1 b_2 - a_2 b_1) \cdot \Delta\left(\vec{e}_1, \vec{e}_2\right) \\
&= (a_1 b_2 - a_2 b_1) \cdot \Delta(I_2),
\end{aligned}$$

where $I_2 = \begin{bmatrix} 1 & 0 \\ 0 & 1 \end{bmatrix}$ is the identity matrix (of order 2), because

$$\Delta(\vec{e}_1, \vec{e}_1) = \Delta(\vec{e}_2, \vec{e}_2) = 0, \quad \text{and} \quad \Delta(\vec{e}_2, \vec{e}_1) = -\Delta(\vec{e}_1, \vec{e}_2).$$

Similar (but much lengthier) computation gives

$$\Delta\begin{bmatrix} a_1 & b_1 & c_1 \\ a_2 & b_2 & c_2 \\ a_3 & b_3 & c_3 \end{bmatrix}$$
$$= (a_1 b_2 c_3 + a_2 b_3 c_1 + a_3 b_1 c_2 - a_3 b_2 c_1 - a_1 b_3 c_2 - a_2 b_1 c_3) \cdot \Delta(I_3),$$

if Δ is an alternate multilinear function of (column) vectors.

It appears that the values of an alternate multilinear function are completely determined by its value at the identity matrix. This is actually the case as we shall see soon.

Theorem 5. Suppose $\Delta(\vec{\alpha}_1, \vec{\alpha}_2, \cdots, \vec{\alpha}_n)$ is an alternate multilinear function of n vectors that is not identically zero. Then

$$\Delta(\vec{\alpha}_1, \vec{\alpha}_2, \cdots, \vec{\alpha}_n) = 0$$

if and only if $\vec{\alpha}_1, \vec{\alpha}_2, \cdots, \vec{\alpha}_n$ are linearly dependent.

Proof. If $\vec{\alpha}_1, \vec{\alpha}_2, \cdots, \vec{\alpha}_n$ are linearly dependent vectors, then there are scalars $k_1, k_2, \cdots, k_n$, not all zero, such that

$$\sum_{j=1}^{n} k_j \vec{\alpha}_j = \vec{0}.$$

Determinants

Suppose $k_i \neq 0$ for some i ($1 \leq i \leq n$), then $\vec{\alpha}_i$ can be expressed as a linear combination of the others:

$$\vec{\alpha}_i = \sum_{j \neq i} c_j \vec{\alpha}_j \quad \left(c_j = -\frac{k_j}{k_i} \right).$$

Therefore,

$$\Delta(\vec{\alpha}_1, \vec{\alpha}_2, \cdots, \vec{\alpha}_n) = \Delta\left(\vec{\alpha}_1, \cdots, \vec{\alpha}_{i-1}, \sum_{j \neq i} c_j \vec{\alpha}_j, \vec{\alpha}_{i+1}, \cdots, \vec{\alpha}_n\right)$$

$$= \sum_{j \neq i} c_j \cdot \Delta(\vec{\alpha}_1, \cdots, \vec{\alpha}_{i-1}, \vec{\alpha}_j, \vec{\alpha}_{i+1}, \cdots, \vec{\alpha}_n) = 0,$$

because each term has two identical vectors.

It remains to show that if $\vec{\alpha}_1, \vec{\alpha}_2, \cdots, \vec{\alpha}_n$ are linearly independent, but

$$\Delta(\vec{\alpha}_1, \vec{\alpha}_2, \cdots, \vec{\alpha}_n) = 0,$$

then $\Delta\left(\vec{\beta}_1, \vec{\beta}_2, \cdots, \vec{\beta}_n\right) = 0$ for an arbitrary set of n vectors $\left\{\vec{\beta}_1, \vec{\beta}_2, \cdots, \vec{\beta}_n\right\}$, i.e., Δ must be identically zero. Because $\{\vec{\alpha}_1, \cdots, \vec{\alpha}_n\}$ forms an ordered basis, we may express each $\vec{\beta}$ as a linear combination of $\vec{\alpha}$'s; i.e.;

$$\vec{\beta}_i = \sum_{j=1}^{n} x_{ij} \vec{\alpha}_j \quad (i = 1, \cdots, n).$$

If we replace each $\vec{\beta}$ in $\Delta\left(\vec{\beta}_1, \cdots, \vec{\beta}_n\right)$ by its expression as a linear combination of $\vec{\alpha}$'s and expand the result by multilinearity (as in the examples above), we obtain a linear combination of terms such as $\Delta(\vec{\gamma}_1, \vec{\gamma}_2, \cdots, \vec{\gamma}_n)$, where each $\vec{\gamma}$ is one of $\vec{\alpha}$'s. If, in such a term, two of the $\vec{\gamma}$'s coincide, then, because Δ is alternate, that term must be zero. So only those terms $\Delta(\vec{\gamma}_1, \vec{\gamma}_2, \cdots, \vec{\gamma}_n)$, where the $\vec{\gamma}$'s are all distinct, hence $\{\vec{\gamma}_1, \cdots, \vec{\gamma}_n\}$ are some permutations of $\{\vec{\alpha}_1, \cdots, \vec{\alpha}_n\}$, will survive. But each permutation can be obtained by repeated interchanges of pairs of vectors, and because Δ is alternate, the sign of its term changes each time we interchange two vectors $\vec{\alpha}_i$ and $\vec{\alpha}_j$. Hence for those $\{\vec{\gamma}_1, \cdots, \vec{\gamma}_n\}$ which are permutations of $\{\vec{\alpha}_1, \cdots, \vec{\alpha}_n\}$, we have

$$\Delta(\vec{\gamma}_1, \cdots, \vec{\gamma}_n) = \pm \Delta(\vec{\alpha}_1, \cdots, \vec{\alpha}_n).$$

More precisely,

$$\Delta(\vec{\gamma}_{\sigma 1}, \cdots, \vec{\gamma}_{\sigma n}) = \operatorname{sgn} \sigma \cdot \Delta(\vec{\alpha}_1, \cdots, \vec{\alpha}_n),$$

where $\operatorname{sgn} \sigma$ of a permutation σ is defined as

$$\operatorname{sgn} \sigma = (-1)^m,$$

and m is the number of interchanges needed to obtain $(1, \cdots, n)$ from $(\sigma 1, \cdots, \sigma n)$. Now, for each permutation σ, there are numerous ways to obtain $(1, \cdots, n)$ from $(\sigma 1, \cdots, \sigma n)$ by repeated interchanges of pairs; however, it turns out that the number of interchanges needed is either always even or always odd. This can be seen from observing the change of sign in the polynomial

$$\begin{aligned} p(t_1, t_2, \cdots, t_n) &= \prod_{i<j}(t_i - t_j) \\ &= (t_1 - t_2)(t_1 - t_3)\cdots(t_1 - t_n) \\ &\quad (t_2 - t_3)\cdots(t_2 - t_n)\cdots(t_{n-1} - t_n) \end{aligned}$$

when the permutation is performed. Thus the permutation is called *even* or *odd*, depending on the number of interchanges of pairs needed to obtain $(1, \cdots, n)$ from $(\sigma 1, \cdots, \sigma n)$. Hence the definition above of the *signum* of a permutation is meaningful.

It follows that

$$\Delta\left(\vec{\beta}_1, \cdots, \vec{\beta}_n\right) = \left\{\sum_\sigma (\text{sgn }\sigma) x_{1(\sigma 1)} x_{2(\sigma 2)} \cdots x_{n(\sigma n)}\right\} \cdot \Delta\left(\vec{\alpha}_1, \vec{\alpha}_2, \cdots, \vec{\alpha}_n\right).$$

Thus if $\Delta(\vec{\alpha}_1, \vec{\alpha}_2, \cdots, \vec{\alpha}_n) = 0$, then $\Delta\left(\vec{\beta}_1, \vec{\beta}_2, \cdots, \vec{\beta}_n\right) = 0$ for every set of n vectors $\left\{\vec{\beta}_1, \vec{\beta}_2, \cdots, \vec{\beta}_n\right\}$, contradicting the assumption that $\Delta \neq 0$. □

Corollary 6. Let Δ be an alternate mutilinear function that is not identically zero. Then a square matrix A is invertible if and only if $\Delta(A) \neq 0$.

Proof. A square matrix is invertible if and only if its (column) vectors are linearly independent. □

Note carefully, the proof of the last theorem furnishes the existence together with the uniqueness (up to a constant multiple) of an alternate multilinear function of n vectors. It shows that the values of an alternate multilinear function of n vectors in the n-space is completely determined by its value on *one* ordered basis. In other words, the set of all alternate multilinear functions of n vectors in the n-space forms a one-dimensional vector space.

Theorem 7. There exists a unique alternate multilinear function Δ of n vectors in the n-space such that

$$\Delta(I_n) = \Delta(\vec{e}_1, \cdots, \vec{e}_n) = 1,$$

where I_n is the identity matrix of order n and $\{\vec{e}_1, \cdots, \vec{e}_n\}$ is the standard basis for the n-space.

Determinants

This function is called the *determinant* and is denoted $\det(\vec{a}_1, \cdots, \vec{a}_n)$, or $\det(A)$, where A is the $n \times n$ matrix whose ith column vector is $\vec{a}_i$ $(1 \leq i \leq n)$.

Recalling that the set of all alternate multilinear functions form a one-dimensional vector space, we obtain the following.

Corollary 8. An alternate multilinear function Δ of n vectors in the n-space must be a constant multiple of the determinant:

$$\Delta(A) = C \cdot \det(A), \quad \text{where } C = \Delta(I_n).$$

Here is an alternate proof. Suppose

$$A = \begin{bmatrix} a_1 & b_1 & \cdots & \ell_1 \\ a_2 & b_2 & \cdots & \ell_2 \\ \cdots & & \cdots & \\ a_n & b_n & \cdots & \ell_n \end{bmatrix}.$$

Because $\Delta(A)$ is multilinear, each term must contain each of $a, b, \cdots, \ell$ once and only once; i.e.,

$$\Delta(A) = \sum C_{(\alpha,\beta,\cdots,\lambda)} a_\alpha b_\beta \cdots \ell_\lambda,$$

where the summation runs through all n-tuples $(\alpha, \beta, \cdots, \lambda)$ of integers 1 through n.

Now to investigate the consequence of Δ being alternate, we choose an arbitrary pair, say a and b, from $a, b, \cdots, \ell$, and denote Δ as follows:

$$\Delta(A) = \sum_{\alpha,\beta=1}^{n} Q_{\alpha\beta} a_\alpha b_\beta,$$

where the Q's contain neither a nor b. Because Δ is alternate, $\Delta = 0$ if $a_\alpha = b_\alpha$ $(\alpha = 1, 2, \cdots, n)$; i.e.,

$$0 = \sum_{\alpha,\beta=1}^{n} Q_{\alpha\beta} a_\alpha a_\beta = \sum_{\alpha=1}^{n} Q_{\alpha\alpha} a_\alpha^2 + \sum_{\alpha \neq \beta}^{n} (Q_{\alpha\beta} + Q_{\beta\alpha}) a_\alpha a_\beta.$$

Because this is true for any $a_1, a_2, \cdots, a_n$, we conclude that $Q_{\alpha\alpha} = 0$ $(\alpha = 1, 2, \cdots, n)$ and $Q_{\alpha\beta} + Q_{\beta\alpha} = 0$ $(\alpha \neq \beta)$.

The conditions $Q_{\alpha\alpha} = 0$ $(\alpha = 1, 2, \cdots, n)$ mean that a term with two letters whose indices are equal, such as $a_\alpha b_\alpha c_\gamma \cdots \ell_\lambda$, should not appear in Δ. Because this must be true for any pair of letters in $a, b, \cdots, \ell$, all the indices in a term must be different; i.e., in each term $\{\alpha, \beta, \cdots, \lambda\}$ is a permutation of $\{1, 2, \cdots, n\}$.

Next, the conditions $Q_{\alpha\beta} + Q_{\beta\alpha} = 0$ $(\alpha \neq \beta)$ mean that two terms which differ by interchanging one pair of indices for a and b, the coefficients of these terms must differ by signs only. This must be true for any pair of letters; hence if Δ contains the term $Ca_1 b_2 c_3 \cdots \ell_n$, then the coefficients of all the terms which can be obtained by interchanging pairs of indices an even number of times must be $+C$, while those with an odd number of times must be $-C$. We conclude that

$$\Delta(A) = C \sum_\sigma (\operatorname{sgn} \sigma) a_{\sigma 1} b_{\sigma 2} \cdots \ell_{\sigma n}.$$

We have shown that if Δ is an alternate multilinear function, then Δ must be of the form above. It is clear that a function of this form is alternate and multilinear. For, the multilinearity obviously holds, while, as for the alternateness, if indices of a pair, say a and b, are interchanged, then the coefficients of the terms

$$a_\alpha b_\beta c_\gamma \cdots \ell_\lambda \quad \text{and} \quad a_\beta b_\alpha c_\gamma \cdots \ell_\lambda$$

differ by sign; hence if $a_\alpha = b_\alpha$ $(\alpha = 1, 2, \cdots, n)$, then the terms cancel pairwise, and we obtain $\Delta = 0$. This is true for all pairs of letters.

The coefficients C can be arbitrary, and the alternate multilinear function Δ of n vectors in the n-space with $C = 1$ is called the *determinant*; i.e.,

$$\det(A) = \sum_\sigma (\operatorname{sgn} \sigma) a_{\sigma 1} b_{\sigma 2} \cdots \ell_{\sigma n}.$$

In particular, $\det(I_n) = 1$, where I_n is the identity matrix of order n.

Thus, from Example 4 (page 227), we obtain

$$\begin{vmatrix} a_1 & b_1 \\ a_2 & b_2 \end{vmatrix} = \det \begin{bmatrix} a_1 & b_1 \\ a_2 & b_2 \end{bmatrix} = a_1 b_2 - a_2 b_1,$$

$$\begin{vmatrix} a_1 & b_1 & c_1 \\ a_2 & b_2 & c_2 \\ a_3 & b_3 & c_3 \end{vmatrix} = \det \begin{bmatrix} a_1 & b_1 & c_1 \\ a_2 & b_2 & c_2 \\ a_3 & b_3 & c_3 \end{bmatrix}$$
$$= a_1 b_2 c_3 - a_2 b_3 c_1 + a_3 b_1 c_2 - a_3 b_2 c_1 - a_1 b_3 c_2 - a_2 b_1 c_3.$$

A.2 Properties

We now discuss some properties of determinants.

Theorem 9. For any two $n \times n$ matrices A and B,

$$\det(AB) = (\det A)(\det B).$$

Proof. Fix the $n \times n$ matrix A, and consider $\Delta(B) = \det(AB)$. We claim that Δ is an alternate multilinear function of matrix B. For, if we denote the ith column vector of matrix B by $\vec{\beta}_i$ $(i = 1, 2, \cdots, n)$, then

$$\Delta(B) = \det \left(A\vec{\beta}_1, A\vec{\beta}_2, \cdots, A\vec{\beta}_n \right),$$

Determinants   233

where $A\vec{\beta}_j$ denotes the column vector obtained as the product of the $n \times n$ matrix A and the $n \times 1$ matrix $\vec{\beta}_j$ $(j = 1, 2, \cdots, n)$. Because

$$A\left(k\vec{\beta}_j + k'\vec{\beta}'_j\right) = k\left(A\vec{\beta}_j\right) + k'\left(A\vec{\beta}'_j\right),$$

and because the determinant is multilinear, it is immediate that $\Delta(B)$ is also multilinear. If $\vec{\beta}_i = \vec{\beta}_j$ $(i \neq j)$, then $A\vec{\beta}_i = A\vec{\beta}_j$, and because the determinant is alternate, we conclude that Δ is also alternate. Therefore, by Corollary 8 (page 231),

$$\begin{aligned} \det(AB) &= \Delta(B) = \Delta(I_n) \cdot \det(B) \\ &= \det(AI_n) \cdot \det(B) = \det(A) \cdot \det(B). \end{aligned}$$

□

Theorem 10. For any square matrix A, $\det(A^t) = \det(A)$, where A^t denotes the transpose of A; i.e., A^t is a matrix whose ith row is the ith column of A.

To prove this theorem, we need some terminology and a lemma. An *elementary matrix* is a square matrix obtained by applying one of the following three operations once (and only once) to the identity matrix I_n:

(a) Interchange two rows.

(b) Multiply a row by a nonzero scalar k.

(c) Add a multiple of a row to another row.

These operations are called *elementary row operations*. Note that the effect of applying an elementary row operation to a (square) matrix is the same as multiplying the corresponding elementary matrix from the left. From this fact and the procedure for computing the inverse matrix of an invertible matrix (by repeated applications of elementary row operations), we obtain that *every invertible matrix can be expressed as a product of elementary matrices*.

Lemma 11. Suppose E is an elementary matrix. Then

(a) $\det(E) = -1$ if E is obtained by interchanging two rows of I_n.

(b) $\det(E) = k$ if E is obtained by multiplying a row of I_n by a nonzero scalar k.

(b) $\det(E) = 1$ if E is obtained by adding a multiple of a row to another row in I_n.

In all cases, $\det(E^t) = \det(E)$.

Proof. (a) The effect of interchanging the ith row and the jth row of I_n is the same as that of interchanging the ith column and the jth column, hence $\det(E) = -1$, because the determinant is alternate. In this case, $E^t = E$, and so $\det(E^t) = \det(E)$.
(b) The effect of multiplying the ith row of I_n by a constant k is the same as that of multiplying the ith column by the same constant k, and so $\det(E) = k \cdot \det(I_n) = k$, by the multilinearity of determinant. Again we have $E^t = E$ in this case; hence $\det(E^t) = \det(E)$.
(c) The effect of multiplying the ith row of I_n by a scalar k, and adding the result to the jth row is the same as that of multiplying the jth column by k and adding the result to the ith column $(i \neq j)$. Hence, by multilinearity,

$$\begin{aligned}\det(E) &= \det(\vec{e}_1, \cdots, \vec{e}_{i-1}, \vec{e}_i + k\vec{e}_j, \vec{e}_{i+1}, \cdots, \vec{e}_n) \\ &= \det(I_n) + k \det(\vec{e}_1, \cdots, \vec{e}_{i-1}, \vec{e}_j, \vec{e}_{i+1}, \cdots, \vec{e}_n) = 1,\end{aligned}$$

because the second term has two identical columns (as $j \neq i$), and the determinant is alternate. In this case, E^t can also be obtained from I_n in the same manner as E was obtained (i.e., by adding a multiple of the ith column to the jth column); hence $\det(E^t) = 1 = \det(E)$. □

Proof of Theorem 10. If A is invertible, then A can be expressed as a product of elementary matrices; $A = E_1 E_2 \cdots E_m$, and so $A^t = E_m^t E_{m-1}^t \cdots E_1^t$. It follows that, by a repeated application of Theorem 9 and Lemma 11,

$$\begin{aligned}\det(A^t) &= \det\left(E_m^t E_{m-1}^t \cdots E_1^t\right) \\ &= \det\left(E_m^t\right) \cdot \det\left(E_{m-1}^t\right) \cdots \det\left(E_1^t\right) \\ &= \det(E_m) \cdot \det(E_{m-1}) \cdots \det(E_1) \\ &= \det(E_1) \cdot \det(E_2) \cdots \det(E_m) \\ &= \det(E_1 E_2 \cdots E_m) = \det(A).\end{aligned}$$

If A is not invertible, then neither is A^t (because "row rank = column rank" for any matrix), and hence we obtain $\det(A^t) = 0 = \det(A)$, by Corollary 6. □

As a consequence, any proposition about determinants which is true for columns is also true for rows, and vice versa. In particular, we have the following.

> **Theorem 12.** (a) If B is obtained by interchanging two rows (two columns) of a square matrix A, then $\det B = -\det A$.
> (b) If B is obtained by multiplying a row (a column) of A by a constant k, then $\det B = k \cdot \det A$.
> (c) If B is obtained by adding a multiple of a row (a column) to another row (column) of A, then $\det B = \det A$.

Alternate Proof. In all three cases, with an appropriate elementary matrix E, we have $B = EA$ for statements about rows, and $B = AE$ in case of columns; hence the result follows from Lemma 11 and Theorem 9.

Determinants

Part (c) of this theorem is employed frequently to compute the values of determinants instead of expanding as in the proof of Theorem 5.

Exercises. (a) Given that 41965, 12801, 73457, 16779, 5457 are all multiples of 17, show that the determinant

$$\begin{vmatrix} 4 & 1 & 9 & 5 & 6 \\ 1 & 2 & 8 & 0 & 1 \\ 7 & 3 & 4 & 5 & 7 \\ 1 & 6 & 7 & 7 & 9 \\ 0 & 5 & 4 & 5 & 7 \end{vmatrix}$$

is also a multiple of 17.

(b) Suppose all the entries, except those on the main diagonal (i.e., the diagonal from the upper left to the lower right), of a determinant are non-positive, but the sum of the entries in each row is positive. Show that the value of the determinant is positive.

A.3 The Laplace Expansion Theorem

Theorem 13. Suppose $\begin{bmatrix} A & B \\ 0 & C \end{bmatrix}$ is an $n \times n$ matrix ($n = r+s$), where A is an $r \times r$ matrix, C is an $s \times s$ matrix, B is an $r \times s$ matrix, and 0 represents the $s \times r$ zero matrix. Then

$$\det \begin{bmatrix} A & B \\ 0 & C \end{bmatrix} = (\det A) \cdot (\det C).$$

Similarly,

$$\det \begin{bmatrix} A & 0 \\ B & C \end{bmatrix} = (\det A) \cdot (\det C),$$

where B is an $s \times r$ matrix and 0 is the $r \times s$ zero matrix in the second equality.

Proof. Set $\Delta(A, B, C) = \det \begin{bmatrix} A & B \\ 0 & C \end{bmatrix}$. If we fix B and C, then Δ is alternate and multilinear as a function of the columns of A. Therefore, by Corollary 8,

$$\Delta(A, B, C) = (\det A) \cdot \Delta(I_r, B, C),$$

where I_r is the $r \times r$ identity matrix. By subtracting multiples of the first r columns of $\begin{bmatrix} I_r & B \\ 0 & C \end{bmatrix}$ (Part (c) of Theorem 12), we obtain

$$\Delta(I_r, B, C) = \Delta(I_r, 0, C).$$

Now $\Delta(I_r, 0, C) = \det \begin{bmatrix} I_r & 0 \\ 0 & C \end{bmatrix}$ is clearly alternate and multilinear as a function of the columns of C. Thus, again by Corollary 8,

$$\Delta(I_r, 0, C) = \Delta(I_r, 0, I_s) \cdot (\det C) = (\det I_n) \cdot (\det C) = \det C,$$

because $\begin{bmatrix} I_r & 0 \\ 0 & I_s \end{bmatrix} = I_n$. So,

$$\det \begin{bmatrix} A & B \\ 0 & C \end{bmatrix} = \Delta(A, B, C) = (\det A) \cdot \Delta(I_r, B, C)$$
$$= (\det A) \cdot \Delta(I_r, 0, C) = (\det A) \cdot (\det C).$$

□

Theorem 14. Suppose $A = \begin{bmatrix} a_{11} & a_{12} & \cdots & a_{1n} \\ a_{21} & a_{22} & \cdots & a_{2n} \\ \cdots & & & \\ a_{n1} & a_{n2} & \cdots & a_{nn} \end{bmatrix}$. Set

$$\Delta_{ij}(A) = \sum_{k=1}^{n} (-1)^{i+k} a_{ik} (\det A_{jk}),$$
$$\Delta^{ij}(A) = \sum_{k=1}^{n} (-1)^{k+i} a_{ki} (\det A_{kj}) \quad (i, j = 1, 2, \cdots, n),$$

where A_{jk} is the $(n-1) \times (n-1)$ matrix obtained by deleting the jth row and kth column of A. Then

$$\Delta_{ij}(A) = \Delta^{ij}(A) = \delta_{ij} \cdot (\det A),$$

where δ_{ij} is the *Kronecker delta*:

$$\delta_{ij} = \begin{cases} 1 & (i = j); \\ 0 & (i \neq j). \end{cases}$$

Proof. It is easy to verify that Δ_{ij} are alternate and multilinear with respect to the column vectors, while Δ^{ij} are alternate and multilinear with respect to the row vectors. Hence all that remains is to evaluate their values at I_n. □

Definition 15. Suppose A is an $n \times n$ matrix. $\det A_{ij}$ is called the *minor* of a_{ij}, and $(-1)^{i+j} \det A_{ij}$ is called the *cofactor* of a_{ij}. Furthermore, the *cofactor matrix* C of A is the $n \times n$ matrix whose (i, j)-entry c_{ij} is the (i, j)-cofactor of a_{ij}. The *adjoint* of A, denoted adj A, is the transpose of this cofactor matrix of C.

Determinants

Thus

$$A \cdot (\operatorname{adj} A) = \begin{bmatrix} a_{11} & a_{12} & \cdots & a_{1n} \\ a_{21} & a_{22} & \cdots & a_{2n} \\ \cdots & \cdots & & \cdots \\ a_{n1} & a_{n2} & \cdots & a_{nn} \end{bmatrix} \cdot \begin{bmatrix} c_{11} & c_{21} & \cdots & c_{n1} \\ c_{12} & a_{22} & \cdots & c_{n2} \\ \cdots & \cdots & & \cdots \\ c_{1n} & c_{2n} & \cdots & c_{nn} \end{bmatrix}$$

$$= \begin{bmatrix} \det A & & & 0 \\ & \det A & & \\ & & \ddots & \\ 0 & & & \det A \end{bmatrix} \quad \text{(by Theorem 14)}$$

$$= (\det A) \cdot I_n$$

Hence, if $\det A \neq 0$, then $A^{-1} = \frac{1}{\det A} \cdot (\operatorname{adj} A)$.

Theorem 16. For any square matrix A,

$$A \cdot (\operatorname{adj} A) = (\det A) \cdot I_n = (\operatorname{adj} A) \cdot A.$$

Moreover, if $\det A \neq 0$, then

$$A^{-1} = \frac{1}{\det A} \cdot (\operatorname{adj} A).$$

Theorem 14 can be extended further. Let A be an $n \times n$ matrix, and $1 \leq r < n$. Choose r rows ($1 \leq i_1 < i_2 < \cdots < i_r \leq n$), and r columns ($1 \leq j_1 < j_2 < \cdots < j_r \leq n$), and form the determinant

$$\Delta \begin{pmatrix} i_1 & i_2 & \cdots & i_r \\ j_1 & j_2 & \cdots & j_r \end{pmatrix}$$

(keeping their orders) at the intersection of these r rows and r columns. On the other hand, we can also consider the determinant obtained by discarding r rows and r columns. The product of the resulting determinant and $(-1)^{\alpha+\beta}$, where $\alpha = i_1 + i_2 + \cdots + i_r$, $\beta = j_1 + j_2 + \cdots + j_r$, is called the *cofactor* of $\Delta \begin{pmatrix} i_1 & i_2 & \cdots & i_r \\ j_1 & j_2 & \cdots & j_r \end{pmatrix}$ and is denoted by

$$\bar{\Delta} \begin{pmatrix} i_1 & i_2 & \cdots & i_r \\ j_1 & j_2 & \cdots & j_r \end{pmatrix}.$$

Theorem 17. [Laplace] Fix $(i_1, i_2, \cdots, i_r)$. Then

$$\det A = \sum_{(j_1, j_2, \cdots, j_r)} \Delta \begin{pmatrix} i_1 & i_2 & \cdots & i_r \\ j_1 & j_2 & \cdots & j_r \end{pmatrix} \cdot \bar{\Delta} \begin{pmatrix} i_1 & i_2 & \cdots & i_r \\ j_1 & j_2 & \cdots & j_r \end{pmatrix}$$

$$= \sum_{(j_1, j_2, \cdots, j_r)} \Delta \begin{pmatrix} j_1 & j_2 & \cdots & j_r \\ i_1 & i_2 & \cdots & i_r \end{pmatrix} \cdot \bar{\Delta} \begin{pmatrix} j_1 & j_2 & \cdots & j_r \\ i_1 & i_2 & \cdots & i_r \end{pmatrix},$$

where $\sum_{(j_1,j_2,\cdots,j_r)}$ runs over all $\binom{n}{r}$ ways of choosing $(j_1, j_2, \cdots, j_r)$. Furthermore, if $(i_1, i_2, \cdots, i_r) \neq (k_1, k_2, \cdots, k_r)$, then

$$\sum_{(j_1,j_2,\cdots,j_r)} \Delta\begin{pmatrix} i_1 & i_2 & \cdots & i_r \\ j_1 & j_2 & \cdots & j_r \end{pmatrix} \cdot \bar{\Delta}\begin{pmatrix} k_1 & k_2 & \cdots & k_r \\ j_1 & j_2 & \cdots & j_r \end{pmatrix}$$

$$= \sum_{(j_1,j_2,\cdots,j_r)} \Delta\begin{pmatrix} j_1 & j_2 & \cdots & j_r \\ i_1 & i_2 & \cdots & i_r \end{pmatrix} \cdot \bar{\Delta}\begin{pmatrix} j_1 & j_2 & \cdots & j_r \\ k_1 & k_2 & \cdots & k_r \end{pmatrix} = 0.$$

Proof. Imitate the proof of Theorem 14. □

Exercise. In the expansion of a determinant of order n, how many terms do not contain any entries from the main diagonal (i.e., the one from the upper left corner to the lower right corner)?

www.ingramcontent.com/pod-product-compliance
Lightning Source LLC
LaVergne TN
LVHW091536060526
838200LV00036B/630